04-HHE-6

FATE OF ESTROGENIC COMPOUNDS DURING MUNICIPAL SLUDGE STABILIZATION AND DEWATERING

by:

Edward T. Furlong
U.S. Geological Survey

James L. Gray
U.S. Geological Survey

David M. Quanrud
University of Arizona

Sondra S. Teske
University of Arizona

Kathleen Esposito
AECOM

Jeremy Marine
AECOM

Wendell P. Ela
University of Arizona

Beverley Stinson
AECOM

Dana W. Kolpin
U.S. Geological Survey

Patrick J. Phillips
U.S. Geological Survey

2010

WERF

The Water Environment Research Foundation, a not-for-profit organization, funds and manages water quality research for its subscribers through a diverse public-private partnership between municipal utilities, corporations, academia, industry, and the federal government. WERF subscribers include municipal and regional water and wastewater utilities, industrial corporations, environmental engineering firms, and others that share a commitment to cost-effective water quality solutions. WERF is dedicated to advancing science and technology addressing water quality issues as they impact water resources, the atmosphere, the lands, and quality of life.

For more information, contact:
Water Environment Research Foundation
635 Slaters Lane, Suite G-110
Alexandria, VA 22314-1177
Tel: (571) 384-2100
Fax: (703) 299-0742
www.werf.org
werf@werf.org

This report was co-published by the following organization.

IWA Publishing
Alliance House, 12 Caxton Street
London SW1H 0QS, United Kingdom
Tel: +44 (0) 20 7654 5500
Fax: +44 (0) 20 7654 5555
www.iwapublishing.com
publications@iwap.co.uk

Library of Congress Catalog Card Number: 2010920053
IWAP ISBN: 978-1-84339-390-0/1-84339-390-5

U.S. Geological Survey, The University of Arizona, AECOM

The research on which this report is based was developed, in part, by the United States Environmental Protection Agency (EPA) through Cooperative Agreement No. CR83155901-1 with the Water Environment Research Foundation (WERF). However, the views expressed in this document are not necessarily those of the EPA and EPA does not endorse any products or commercial services mentioned in this publication. This report is a publication of WERF, not EPA. Funds awarded under the Cooperative Agreement cited above were not used for editorial services, reproduction, printing, or distribution.

ACKNOWLEDGMENTS

The research team would like to thank the wastewater treatment plants that agreed to participate in this study. The assistance provided by plant personnel in sample collection and provision of operating data was very much appreciated and was critical to the success of this project. The team is grateful for the sample collection efforts of Daniel Edwards of the U.S. Geological Survey and Marija Peric of AECOM as well as the technical review of plant flow and solids load data by Dr. Mohammad Abu-Orf and Dr, Gregory Bowden, both of AECOM. The team would also like to thank the members of the Project Subcommittee, listed below, for their thorough reviews and support. Lastly, the team would like to thank the WERF Program Manager, Alan Hais, for his commitment to and enthusiasm for the importance of this project.

Research Team

Principal Investigators:
Edward T. Furlong, Ph.D.
U.S. Geological Survey

David M. Quanrud, Ph.D.
The University of Arizona

Beverley Stinson, Ph.D.
AECOM

Project Team:
Wendell P. Ela, Ph.D.
University of Arizona

Kathleen Esposito, LEED AP
AECOM

James L. Gray, Ph.D.
U.S. Geological Survey

Dana W. Kolpin
U.S. Geological Survey

Jeremy Marine
AECOM

Patrick J. Phillips
U.S. Geological Survey

Sondra S. Teske, Ph.D.
University of Arizona

WERF Project Subcommittee

Bob Arnold, Ph.D., *Research Committee Liaison*
University of Arizona

Richard David Holbrook, Jr., Ph.D., P.E.
National Institute of Standards and Technology

Frederic Leusch, Ph.D.
National Research Centre for Environmental Toxicology (EnTox)

Sudhir N. Murthy, Ph.D., P.E.
D.C. Water and Sewer Authority, Department of Wastewater Treatment

Clifford P. Rice
U.S. Department of Agriculture, Agricultural Research Service Environmental Quality Lab

Gregory Sayles
U. S. Environmental Protection Agency

Chi-Chung Tang
L.A. County Sanitation District

Water Environment Research Foundation Staff

Director of Research: Daniel M. Woltering, Ph.D.

Program Directors: Alan Hais, P.E.

ABSTRACT AND BENEFITS

Abstract

This project convened a team of experts in the fields of environmental engineering (AECOM), analytical chemistry and hydrogeology (USGS), and biological assay analysis (UA) to evaluate the occurrence and fate of estrogenic compounds, and the estrogenicity of biosolids derived from wastewater treatment.

Sludge and biosolids samples were collected through the solids treatment train of four wastewater treatment plants (WWTPs) operating a range of solids processing, treatment and disposal options that are typical to facilities across the United States. Targeted solids processing methods included thickening via gravity, gravity belt, and dissolved air flotation; stabilization via lime addition, aerobic digestion and anaerobic digestion; chemical conditioning; dewatering via centrifuge; and other processes including composting and pelletization. Targeted disposal options included beneficial reuse or disposal including land application, dedicated land disposal, and landfilling.

Samples were collected from the study plants between two and five times over two years, allowing for a preliminary assessment of seasonal and annual variation. In some cases, sampling density was not sufficient to assess seasonal variations, but for certain compounds interesting seasonal trends were observed. The solids samples were complimented with liquid samples at key locations in the study plants during several sample collection events. Over the course of the study, 15 sample trips were conducted and a total of 90 samples were collected from the four study plants.

For each sample collected, chemical analysis for 19 steroid hormones and *in vitro* biological assay (bioassay) measurements were conducted to quantify estrogen receptor agonists and estrogenic activity. In addition to the estrogenic compounds, samples were analyzed for a suite of trace organic compounds (TOrCs including anthropogenic wastewater indicators (AWIs) and pharmaceuticals, resulting in analysis for 100 chemical compounds in each liquid or solid sample. Collection of these data substantially expanded the scope and value of the study, providing a more comprehensive evaluation of the effects of wastewater treatment, with specific emphasis on solids processing, on TOrCs.

Loads of TOrCs and estrogenic activity were calculated for each sample point based on flows and solids loadings data from the study plants. In this exercise, TOrC concentrations were multiplied by the solids loading (tons per day) to calculate the daily load of each compound in grams per day (g/day).

This report provides comparisons of the chemical and biological assays used in this study, the results of select TOrC mass balances as well as a discussion of the results and areas for future research.

Benefits

- Provides insight into the primary sources of estrogenic activity in biosolids through comparison of estrogen analyses and measures of whole-sample estrogenic activity.
- Provides much needed information on the occurrence, concentration, characteristics, seasonal variation and potency of estrogenic compounds that are predicted to preferentially partition onto biosolids during common wastewater treatment processes.
- Provides an important step for developing information critical to the assessment of the potential risks associated with biosolids disposal on land.

Keywords: Wastewater treatment, biosolids, cstrogenicity, bioassay, YES, KBluc, trace organics, chemical analysis, hormones, steroids, pharmaceuticals, anthropogenic wastewater indicators.

TABLE OF CONTENTS

LIST OF TABLES

LIST OF FIGURES

LIST OF ACRONYMS

APEO	Alkylphenol ethoxylates
ASE	Accelerated solvent extraction
AWI	Anthropogenic Wastewater Indicator
BOD	Biochemical Oxygen Demand
BPA	Bisphenol A
CASRN	Chemical Abstracts Service Registration Number
CCV	Continuing calibration verification
CF	Concentration factor
CO_2	Carbon dioxide
CPRG	Chlorophenol red β-D-galactopyranoside
DEHP	Diethylhexyl phthalate
DES	Diethylstilbestrol
E1	Estrone
E2	17-beta-estradiol
E2	Estradiol
E3	Estriol
EC	Emerging contaminants
EDCs	Endocrine disrupting compounds
EE2	17α-ethinyl estradiol
EE2-Eqs	EE2-equivalents
ESI	Electrospray ionization
FBS	Fetal bovine serum
FR	First response
g	Gram
GC/MS	Gas chromatography with mass spectrometry
GC/MS/MS	Gas chromatography with tandem mass spectrometry
GPM	Gallons per minute
HPLC	High-performance liquid chromatography
IC	Ion chromatography
IPA	Isopropyl alcohol
K_{OW}	Octanol-water partitioning coefficients
L	Liter
LC/MS	Liquid chromatography with mass spectrometry
LS/MS/MS	Liquid chromatography with tandem mass spectrometry
MAE	Microwave assisted extraction
MDS	Microwave digestion system
MeOH	Methanol
μg	Microgram
μL	Microliter
MGD	Million gallons per day
mL	milliliter
MS	Mass spectrometer
ng	Nanogram
NP	Nonylphenol

NP1EO	Nonylphenol Monoethoxylate
NP2EO	Nonylphenol diethoxylate
NWIS	National Water Information System
NWQL	National Water Quality Laboratory
OP	Octylphenol
OP1EO	Octylphenol monoethoxylate
OP2EO	Octylphenol diethoxylate
PEF	Process Evaluation Facility
ppb	Parts per billion (= micrograms per liter)
ppt	Parts per trillion (= nanograms per liter)
psig	Pounds per square inch gauge
QA	Quality Assurance
QC	Quality Control
RCF	Relative centrifugal force
RLU	Relative light units
SIM	Selected-ion monitoring
SLD	Significance level determination
SPE	Solid-phase extraction
SRT	Solids retention time
TOrC	Trace Organic Compound
TSS	Total Suspended Solids
TWAS	Thickened waste activated sludge
UA	University of Arizona
U.S. EPA	United States Environmental Protection Agency
USGS	United States Geological Survey
VSS	Volatile Suspended Solids
WAS	Waste Activated Sludge
WWTP	Wastewater treatment plant
YES	Yeast estrogen screen

EXECUTIVE SUMMARY

To date, most research on the occurrence, fate, and transport of trace organic compounds (TOrCs) in wastewater treatment plants (WWTPs) has focused on the liquid phase of the treatment train. This is in part due to connections being established between effluent discharges from WWTPs and endocrine disruption in aquatic organisms. Similar to WWTP effluents, biosolids are a potential source of TOrCs to the environment in their frequent and growing use in landscaping, land reclamation and agriculture and additionally to surface water via runoff and groundwater via infiltration.

Biosolids are the largest by-product resulting from wastewater treatment processes. Federal and state regulatory agencies generally encourage the practice of biosolids disposal via addition to soil, and the end use of biosolids is stringently regulated in the United States. However, as more TOrCs are identified and public interest increases, research is necessary to better understand and communicate the implications of the occurrence of TOrCs in biosolids. To conduct this research, it is also necessary to develop analytical techniques capable of measuring trace levels of these compounds in complex matrices such as biosolids. This project convened a team of experts in the fields of environmental engineering (AECOM), analytical chemistry and hydrogeology (U.S. Geological Survey), and biological assay analysis (University of Arizona) to evaluate the occurrence and fate of estrogenic TOrCs, and the total estrogenicity of biosolids derived from wastewater treatment.

This was the first research on TOrCs in sludge and biosolids supported by WERF. The primary objective of this study was to provide key baseline information concerning the estrogenicity (measured with *in vitro* bioassays) and concentrations of individual estrogenic TOrCs (measured using gas chromatography with mass spectrometry (GC/MS/MS) and liquid chromatography with mass spectrometry (LC/MS) methods) through common wastewater treatment processes. These include secondary treatment with activated sludge and processes meant to condition, thicken, stabilize, and dewater sludge. Secondary objectives of this study included: calculation of a mass balance of known endocrine disrupting compounds (EDCs); analysis for other non-estrogenic TOrCs of interest (e.g. carbamazepine, a pharmaceutical, and the insect repellent DEET); assessment of seasonal and annual variation in loads and removal of TOrCs; evaluation of bioassays to analyze complex solids samples; and analysis of the correlation between bio- and chemical assays for both liquids and solids samples.

Four WWTPs across the United States participated in this study. These plants operate a range of sludge and biosolids treatment processes, including: thickening via gravity, gravity belt, and dissolved air flotation; stabilization via lime addition, aerobic digestion and anaerobic digestion (thermophilic and mesophilic); chemical conditioning; and dewatering via centrifuge; and other processes including composting and pelletization. Sample locations were established through the solids, and in some cases liquid, treatment train of each of the study plants and collection, in accordance with USGS (United States Geological Survey) protocols, occurred two to four times over one year resulting in 15 samples trips and a total of 90 samples collected for analysis. Although one sample per plant per season did not prove sufficient to conduct a full analysis of seasonal variability, certain seasonal trends were apparent in plant performance.

For each sample collected, chemical analysis for 19 steroid hormones and *in vitro* bioassay measurements were conducted. The USGS National Water Quality Laboratory (NWQL) provided chemical analysis of samples. In additional to the method to detect 19 natural and synthetic hormones in both liquid and solids samples, methods to identify a wide range of other TOrCs including pharmaceuticals and synthetic organic EDCs also were used, resulting in 100 analytes assayed in solid and liquid media.

The University of Arizona (UA) used the yeast estrogen screen (YES) bioassay to evaluate estrogen agonist/antagonist activity in samples. For a select subset of samples, the relatively newer T47D-KBluc (KBluc) bioassay complimented YES analysis. It is known that different bioassays respond differently to particular estrogenic compounds and in different water matrices. Consequently the utilization of a second bioassay broadened the information gleaned from the sampling effort and provided further cross-comparison between bioassay and single compound methods of quantifying a sample's estrogenic signal. The traditional technique to quantify estrogenic activity in environmental samples relies upon identifying the midpoint (50%, EC50) level of response in both the environmental sample and positive control (either 17-beta-estradiol (E2) or 17α-ethinyl estradiol (EE2)) dose response curves. In this project, estrogenic activity in several samples could not be determined using the traditional EC50 method due to sample toxicity inhibiting the estrogenic response and a new data reduction method, the "First Response" method was devised.

To supplement results, the mass fluxes at each sampling point were predicted using the Model of Concentration Addition, which simply assumes that the estrogenic contribution of each individual compound is linearly additive, so that a summation of all compound's concentration multiplied by their respective EE2-equivalent potency factors is the expected total estrogenicity of the sample (in EE2 equivalents). The results calculated using the Model of Concentration Addition were compared with bioassay results.

Due to the extensive amount of data generated in this study, analytical results for both chemical and bioassay analysis were compiled and published as a separate USGS Data Report (Furlong et al., 2010) that is available on the USGS website (http://pubs.er.usgs.gov/).

To calculate the instantaneous load of TOrCs and estrogenic activity, analytical results were multiplied by the solid loading for each sample point (e.g. tons per day) to obtain the daily load of each compound, presented in grams per day (g/day). In cases where removal or production of a component was significant, conclusions were drawn regarding removal. However, due to the uncertainties described above, the precision of these estimates is somewhat compromised.

The 90 samples, many of which were separated for analysis of both liquid and solid phases due to the high liquid fraction of many sludge streams (e.g. unthickened sludge), were analyzed using both chemical and biological assays for this project. These data were integrated with flow and solids data from four large WWTPs to calculate instantaneous loads of both individual TOrCs and estrogenicity through the treatment trains. The resulting data set was large and complex. There was a high degree of variability in both plant data and analytical results between sampling dates and within and among the plants, complicating the ability to make conclusive interpretations. Furthermore, concentrations of TOrCs were often at or near analytical limits of detection. Lastly, the complexity of solids matrices also must be considered. When interpreting these results, it is important to note that failure to detect a compound in a particular sample does not necessarily imply its absence, simply that its concentration was below the detection limit.

The research team committed to making the best possible interpretations based on the sometimes ambiguous results obtained for each plant. These interpretations are presented in the Results and Discussion section of this report.

Results and corollary discussion are organized into three major groups. First, the general discussion of the chemical and biological data reduction approaches is presented. This is followed by presentation and discussion of results for each plant, including: calculated instantaneous loads for hormones, alkylphenolic compounds, and bioassays; chemical analysis data reduction results and discussion; followed by discussion of the data reduction results of biological analysis and the Model of Concentration Addition approach. The final subsection discusses non-estrogenic TOrCs (e.g. select pharmaceuticals).

Concentrations of estrogenic, androgenic, and other endocrine-active hormones, synthetic estrogenic compounds, pharmaceuticals and other TOrCs were determined in solids and liquids in the four plants studied. These concentrations were then used to estimate instantaneous loads and percent (%) decreases for unit processes in the four plants. Results for hormones and synthetic estrogenic compounds in solids were emphasized in the sampling strategy and resulting data, as potential reduction in estrogenicity in solids through treatment was the primary focus of the study. The load of many TOrCs, particularly the steroid hormones, was decreased very efficiently by secondary treatment processes. One consequence of this is that some concentrations were very close to analytical limits of detection in solid samples. As noted above, this added additional uncertainty to the measurements and complicates the interpretation of data for solids unit processes. In many cases, no evaluation of removal of a specific compound was possible because it was not detected in solid samples.

Within unit processes, particularly during secondary treatment, estrogens and synthetic EDCs undergo phase transfer, transformation to intermediates of differing estrogenic potency, and removals that are unit process specific, with increases in some compounds, such as lower homologue alkylphenol ethoxylates and in some cases estrone. Concurrently, changes in constituent potency via transformation, results in an overall qualitative correspondence to the observed reduction in potency reflected in bioassay estrogenicity. Careful consideration of the exact chemical compositions of liquids and solids moving through treatment is critical to determining which unit processes are most effective at reduction of estrogenicity, and how unit processes may be modified or optimized for maximal reduction in total estrogenicity of solids.

Pharmaceutical loadings also exhibited an array of behaviors in the two plants where samples and results from liquid and solid phases was a sampling design focus. Some compounds, such as carbamazepine, appear minimally removed from the liquid phase or chemically transformed during treatment. The loads of other compounds, such as caffeine, were effectively decreased during treatment. Seasonality may play a substantial role in removal efficiency, although future research with more focused and frequent sampling of specific processes is necessary to better elucidate these effects.

In comparing the predicted and measured results for total estrogenicity, , both bioassays used (YES and KBluc) indicated a lower estrogenicity in the samples than those calculated based on the individual compound concentrations and potencies using the Model of Concentration Addition. This discrepancy is not unexpected as the individual compound approach neglects biological process issues such as competitive binding by different compounds for the estrogen receptor sites, diminished transport of agonists into the bioassay cells due to wastewater matrix effects, and the role of estrogenic antagonists in the wastewater matrix. The KBluc results were typically somewhat higher than the YES results, but still less than the individual compound predicted values. Lastly, the measured (bioassay) vs. predicted (Model of Concentration Addition) results for liquid data showed better agreement than those for solids.

Out of the suite of 100 compounds measured and based on the Model of Concentration Addition, nearly all of the estrogenicity in all plants and all dates was due to the presence of 16 TOrCs, namely the steroidal compounds (mainly estrone and estradiol) and the alkylphenols (mainly nonylphenol and short chain ethoxylates).

Activated sludge treatment (an aerobic process) of the primary effluent substantially decreased estrogenicity. More than 90% of most estrogenic TOrCs are removed from the liquid phase during activated sludge treatment and most of the total estrogenicity in liquids was due to steroidal hormones. According to bioassay analysis, the only biosolids stabilization process that reduced estrogenicity was aerobic digestion, in which a modest 18% reduction was observed. Lime addition resulted in an increase in estrogenicity (although the load of individual estrogenic compounds decreased) in the biosolids, whereas mesophilic and thermophilic anaerobic digestion caused less significant increases. The increase in estrogenicity during anaerobic digestion processes was a consequence of an increased contribution by alkylphenols, particularly nonylphenol (NP), which is more estrogenically potent than its ethoxylated precursors. NP is largely removed during aerobic processes.

For the plants employing anaerobic digestion, the total estrogenicity leaving the plant in the biosolids was greater than that leaving the plant in the secondary treated effluent; although for the two plants in which plant level estrogenic instantaneous load balances could be evaluated, the estrogenicity leaving the plant (liquids plus biosolids) was less than that entering (primary influent).

There are many research needs that emerged in the process of conducting this study, including evaluation of: digestion processes; the fate of TOrCs in biosolids following land application; the effect of lime stabilization on TOrCs; the effect of polymer on transport of TOrCs, particularly in centrate streams; and a more controlled study on the compatibility of chemical analysis and bioassays. In any future research efforts, particularly those conducted at a plant scale, a focus on high frequency sampling is necessary to better capture process variability, particularly when attempting to elucidate seasonal effects.

CHAPTER 1.0

INTRODUCTION

1.1 Background

The presence of a wide array of chemical compounds that are commonly used in commerce including: prescription and non-prescription pharmaceuticals, personal care products, flame retardants, antimicrobials, detergents, pesticides, and natural and synthetic hormones in the environment has been widely documented in recent years (Kolpin, 2002; Glassmeyer, 2008). The scientific community has not reached consensus on an appropriate term for these compounds, which have been referred to using terms such as: emerging contaminants, microconstituents of potential concern, and trace organic compounds (TOrCs).

A subset of TOrCs are known or suspected endocrine disrupting compounds (EDCs); (Sumpter, 2005), which are both naturally occurring and synthetic organic compounds that have the ability to alter the normal function of the endocrine system, which is responsible for growth and development in vertebrates. To date, the term EDC typically refers to compounds that modulate estrogen receptors, resulting in abnormal sexual characteristics such as intersex, atypical male:female sex ratios and other potentially deleterious reproductive effects observed in fish exposed to these compounds (Vajda et al., 2008). However, the realm of potential EDCs includes compounds that could interfere with numerous endocrine axes by multiple mechanisms (i.e., in addition to receptor binding). The EDCs best known to produce these specifically estrogenic effects are the naturally occurring steroidal estrogens, including 17-β-Estradiol, synthetic estrogens, such as ethynyl estradiol, used in birth control, and synthetic organic compounds that have been shown to interact with estrogen receptors, including the alkylphenol ethoxylates, bisphenol A, and a number of phthalate plasticizers. In this context, these EDCs produce the "estrogenicity" of a liquid or solid environmental sample, quantified using a purified standardized receptor bioassay.

WWTPs have been identified as a primary source of TOrCs to water resources as a function of the waste streams they collect (Johnson and Sumpter, 2001; Snyder et al., 2003). Connections are being made between WWTP discharges and endocrine disruption in aquatic organisms. This was recently documented in Colorado where a strong correlation between sexual disruption in fish and environmentally relevant concentrations of TOrCs associated with a WWTP effluent discharge was shown (Vajda et al., 2008). Thus WWTPs can be a critical control point for the mitigation of TOrCs in the environment.

To date, most research on the occurrence, fate and transport of TOrCs in WWTPs has focused on the liquid phase of the wastewater treatment train. This is in part due to the above-referenced link between effluent discharges and endocrine disruption in aquatic organisms. It is also due to the difficulty associated with analyzing solids samples. Similar to WWTP effluents, biosolids are a potential source of TOrCs to the environment (Kinney et al., 2006) in their frequent and growing use in landscaping, land reclamation and agriculture and additionally to surface water via runoff and groundwater via infiltration.

In general, removal of TOrCs from the aqueous phase is not well characterized and processes that mediate removal, such as chemical or biological transformation, or removal by physical means (e.g. sorption to solids) requires additional research (Liu et al., 2009). The hydrophobic property of known estrogenic compounds in wastewater suggests that they may be strongly associated with sludges derived from wastewater treatment. For instance, alkylphenol polyethoxylates, a class of surfactants known to be highly estrogenic, are reported to degrade during the biological activated sludge process to produce estrogenic metabolites (e.g. alkylphenols, alkylphenol monoethoxylates and alkylphenol diethoxylates, and alkylphenol ethoxycarboxylates). While these compounds may not be more persistent or biologically disruptive than the parent compounds, incomplete degradation of the parent compounds may provide an ongoing source of material that can be degraded to more active metabolites. Ahel et al. (1994) reported that while alkylphenol surfactants can be efficiently removed or altered during aerobic treatment, their metabolites have a high octanol water partitioning co-efficient ($K_{ow} > 4.5$) which indicates a preference for sorption to the organically rich waste sludge. However the metabolites are not degraded during anaerobic sludge digestion and tend to accumulate in biosolids. Nevertheless, only a few studies have addressed the fate of EDCs during wastewater treatment and, for chemicals that separate with the sludge, survival during solids handling and treatment processes.

Biosolids are the largest by-product resulting from wastewater treatment processes. Federal and state regulatory agencies have generally encouraged the practice of biosolids disposal via addition to soil (U.S. EPA 1981, 1984, 1991). Nationwide trends in sludge/biosolids disposal reflect increased reliance on the use of biosolids as soil amendments. In year 2001, 68% of the 8,650 publicly owned treatment works that generated sewage sludge in the United States disposed of biosolids via land application or distribution to the public for use as a soil amendment (National Research Council, 2002). This amounts to 3.4×10^6 dry tons of biosolids each year, or 44% of the sewage sludge now produced (U.S. EPA, 1999).

The disposal of wastewater biosolids is stringently regulated in the United States. The U.S. EPA 503 regulations present procedures for the treatment and disposal of wastewater biosolids. Based upon compliance with these standards, the beneficial application of wastewater sludge to agricultural land or disposal to landfills is considered to be fit for their purpose and an environmentally acceptable means of their management. However, as more TOrCs are identified and public interest increases, analytical method development and research is necessary to better understand and communicate the implications of the occurrence of TOrCs in biosolids.

1.2 Study Overview

This report presents the results and findings of an investigation into the fate of known estrogenic compounds, a wide range of other TOrCs and total estrogenic activity in solids, and in some cases liquids, derived from wastewater treatment.

In order to fulfill the objectives of this study it was necessary to:

- Identify sludge and biosolids treatment processes of interest;
- Identify wastewater treatment plants operating the processes of interest to participate;
- Determine sample points and frequency of sample collection;

- Identify target analytes;
- Select analytical method(s) for detection of chemical concentrations; and
- Select bioassay(s) for measurement of estrogenic activity.

Figure 1-1 shows a generalized diagram of solids processing options and Table 1-1 shows the functions of solids processing methods. Of these methods and options, the following were selected for investigation in this study: thickening via gravity, gravity belt, and dissolved air flotation; stabilization via lime addition, aerobic digestion and anaerobic digestion (thermophilic and mesophilic); chemical conditioning; and dewatering via centrifuge; and other processes including composting and pelletization. Targeted disposal options included beneficial reuse or disposal including land application, dedicated land disposal and landfilling.

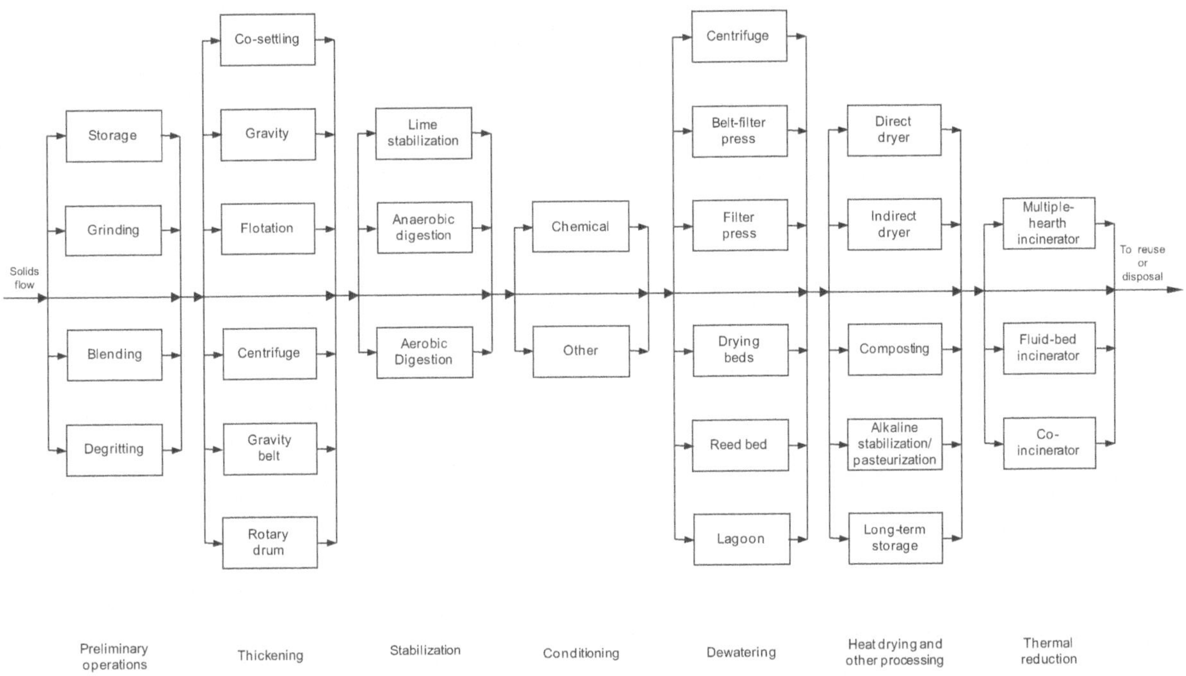

Figure 1-1. Generalized Sludge Processing Flow Diagram (Metcalf and Eddy, 2003).

Table 1-1. Solids Processing Methods.

Unit Operation/Process	Function
Thickening	Volume reduction
Alkaline stabilization	Stabilization
Digestion	Stabilization & mass reduction
Conditioning	Improve dewaterability
Dewatering	Volume reduction

Four WWTPs participated in this study. In addition to operating the target sludge and biosolids treatment processes, the study plants also operate a range of liquid treatment processes that might have an effect on removals to the solids treatment train, such as activated sludge. Sample locations were established through the solids treatment train of each of the study plants. As part of a more comprehensive evaluation, the project team also collected samples from the liquid wastewater treatment streams at two of the study plants. Seasonal impacts were accounted for by repeating sample collection two to four times over one year as shown in Table 1-2. In summary, over the course of this study, 15 samples trips were completed and a total of 90 samples were collected from four WWTPs.

Table 1-2. Sample Collection Trips.

Site	Date
Plant A	3/17/2006
	7/18/2006
	10/4/2006
	1/10/2007
Plant B	12/7/2005
	4/11/2006
	7/18/2006
	10/16/2006
	1/29/2007
Plant C	12/6/2005
	7/17/2006
Plant D	3/16/2006
	6/20/2006
	9/14/2006
	12/4/2006

For each sample collected, chemical analysis for steroid hormones and *in vitro* bioassay measurements were conducted to quantify estrogen receptor agonists and estrogenic activity. In addition to the estrogenic compounds, a substantial subset of samples were analyzed for a suite of anthropogenic wastewater indicators (AWIs) and pharmaceuticals, resulting in analysis for 100 TOrCs in each liquid or solid sample.

The analytical methods used by the USGS are listed in Table 1-3. The analytical methods used for AWIs in water (SH1433), AWIs in sediments/solids (SH5433), pharmaceuticals in water (SH2080) and pharmaceuticals in sediment/solids (LC9008) are described in Zaugg et al. (2002), Burkhardt et al. (2006), Chaill et al. (2004), and Furlong et al. (2008), respectively. The reports detailing the methods for hormones in water (SH4434) and hormones in sediment/solids (SH6434) are still in preparation (Gray et al., *in preparation).* Collection of these data substantially expanded the original scope and value of the study, providing a more comprehensive evaluation of the effects of solids processing and treatment on TOrCs.

Table 1-3. USGS Analytical Methods.

USGS Analytical Method	Schedule/Lab Code	Matrix	# Compounds
Anthropogenic Wastewater Indicators	SH1433	water	58
	SH5433	sediment/solids	59
Pharmaceuticals	SH2080	water	23
	LC9008	sediment/solids	24
Hormones	SH4434	water	19
	SH6434	sediment/solids	19

The UA used two biological assays (bioassays) to measure the estrogenic potency of the samples: the yeast estrogen screen (YES) bioassay and the T47D-KBluc (KBluc) bioassay. Estrogen agonist activities were evaluated in extracts from liquid and solid samples using the YES bioassay.

Operating information, including flows and solids loadings data for each plant was collected. These data were used to calculate the mass balance, or instantaneous loads, of TOrCs and estrogenic activity for each sample point.

1.3 Study Objectives

The primary objective of this study was to provide key baseline information concerning the estrogenicity (measured with *in vitro* bioassays) and concentrations of individual estrogenic TOrCs (measured with chemical analysis) through common wastewater treatment processes to condition, thicken, stabilize, and process sludge.

Secondary objectives of this study included:

- Calculation of instantaneous loads of estrogenicity and comparing them to the expected estrogenicity of the summed estrogenic compounds in liquids and solids
- Analysis for other TOrCs of interest, including pharmaceuticals, personal care products and alkylphenols, resulting in analysis for 100 compounds
- Collection and analysis of liquid samples to compliment solids samples to provide a more accurate picture of the loads of estrogenicity and estrogenic compounds throughout the WWTPs
- Assessment of seasonal variation
- Evaluation of the efficacy of bioassays to analyze complex solids samples
- Analysis of the correlation between bio- and chemical assays for both liquids and solids samples.

CHAPTER 2.0

MATERIALS AND METHODS

2.1 Study Sites

Four WWTPs participated in this study. All are in the United States, with two located on the East Coast, one in the Southwest, and another on the West Coast. The WWTPs will remain anonymous and are identified throughout this report as Plants A, B, C, and D. As stated in the previous section, these plants operate a range of sludge and biosolids treatment processes commonly used to thicken, condition, dewater, reduce and stabilize sludge to produce biosolids. While the solids process trains varied, one commonality across the study plants is that each uses activated sludge for secondary treatment of the liquid stream. As described in this section, of particular interest was the stabilization process at each plant: Plant A uses aerobic digestion; Plant B uses mesophilic anaerobic digestion with both conventional and egg-shaped digesters as well as a two-stage acid-phase digestion process, all operating in parallel trains; Plant C uses lime addition; and Plant D uses thermophilic anaerobic digestion.

This section provides brief descriptions of the WWTPs, simplified process flow schematics that indicate sample collection points, Tables that provide the name and unique station identification assigned by the project team for this study as well as frequency of collection for each sample point.

2.1.1 Plant A

Plant A treats an average flow of 3 MGD (11,356 m^3/d). Plant A utilizes activated sludge secondary treatment with biological nutrient removal (consistently denitrifying) employing oxic and anoxic zones. This plant was specifically selected to participate in this research program because of the mesophilic aerobic digestion process it operates. Following secondary treatment, waste activated sludge is processed by dissolved air flotation, aerobic digestion and centrifuge dewatering. Finished biosolids are currently disposed of by landfilling. A schematic and listing of the sample points for Plant A are provided in Figure 2-1 and Table 2-1, respectively.

Samples were collected from Plant A on March 17, 2006, July 18, 2006, October 4, 2006 and January 1, 2007.

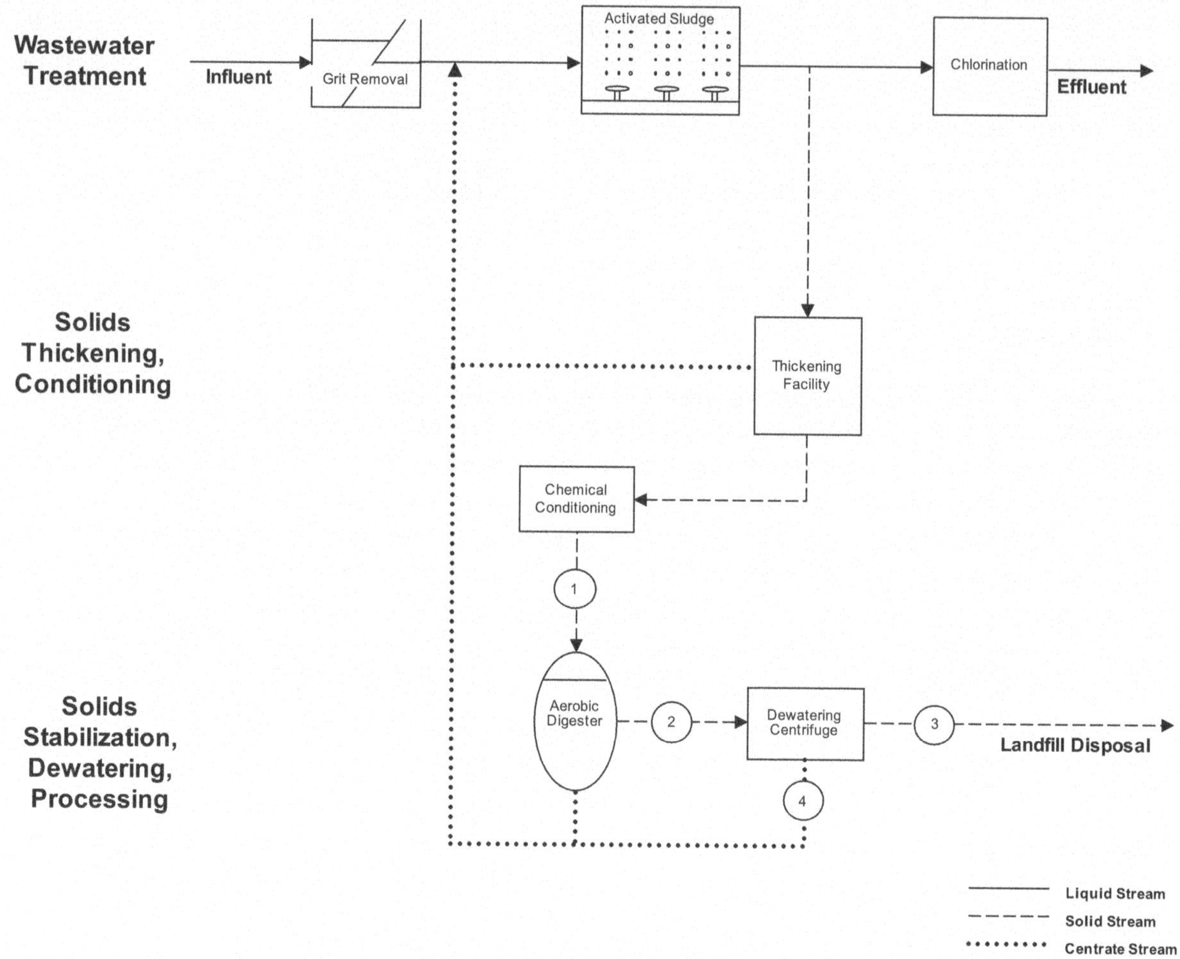

Figure 2-1. Plant A Process Train Schematic.

Table 2-1. Plant A Sampling Points and Frequency.

Figure ID	Location	Sampling Frequency
1	Thickened Sludge (Primary & Secondary)	4x
2	Digested Sludge	4x
3	Dewatered Sludge	4x
4	Centrate Recycle Stream from Dewatering Process	4x

2.1.2 Plant B

Plant B treats an average flow of 155 MGD (586,737 m^3/d). Plant B utilizes fine bubble activated sludge with phosphorus control by chemical addition and nitrogen control by biological processes. Plant B produces an annual average of 31,000 dry tons (28,122,727 kg) per year of biosolids.

Plant B produces biosolids characterized as low in metals and relatively high in nutrients. Plant B was selected because of the diverse range of biosolids treatment processes operated at the facility often in parallel process trains. For example, the plant operates two parallel biosolids digestion processes, a conventional mesophilic anaerobic digestion process and an innovative two-stage acid phase digestion process. Biosolids from Plant B are recycled using four methods: direct agricultural land application, composting of digested biosolids cake for marketing as a soil amendment and fertilizer, heat drying of digested biosolids to produce a dry pelletized product, which is also marketed as a fertilizer, and occasional landfilling when weather conditions prevent land application. Collectively these four biosolids utilization options represent 22, 25, 50, and 3% of the plant's biosolids management program, respectively.

A schematic and list of the sample points for Plant B are provided in Figure 2-2 and Table 2-2, respectively. A preliminary sample collection trip was conducted on December 7, 2005 at Plant B. This trip was to refine sample collection protocols, shipping and handling procedures, and chain of custody forms. Samples were collected from points 6-13. Sample point 7 was not included in the original scope of work, but was added for this collection period to compare anaerobic digestion in the plant's newer egg-shaped anaerobic digesters (sample point 8) and the plant's older conventional anaerobic digesters.

Samples 1-5 and 8-13 were collected from Plant B on July 18, 2006 and January 30, 2007. During the January 2007 sampling trip the Process Evaluation Facility (PEF), which houses the Two-Stage Acid Phase Digestion process, was out of operation. Samples were collected from points 9 and 10 on June 7, 2007, when the PEF was operating at steady state. On the April 11, 2006 and October 16, 2006 sample collection trips, samples were collected from points 6, 8, 11, and 12. It was difficult to obtain a sample of the composted sludge (sample point 14) since the compost is produced at an offsite facility, only one set of samples were collected of composted sludge on February 28, 2007.

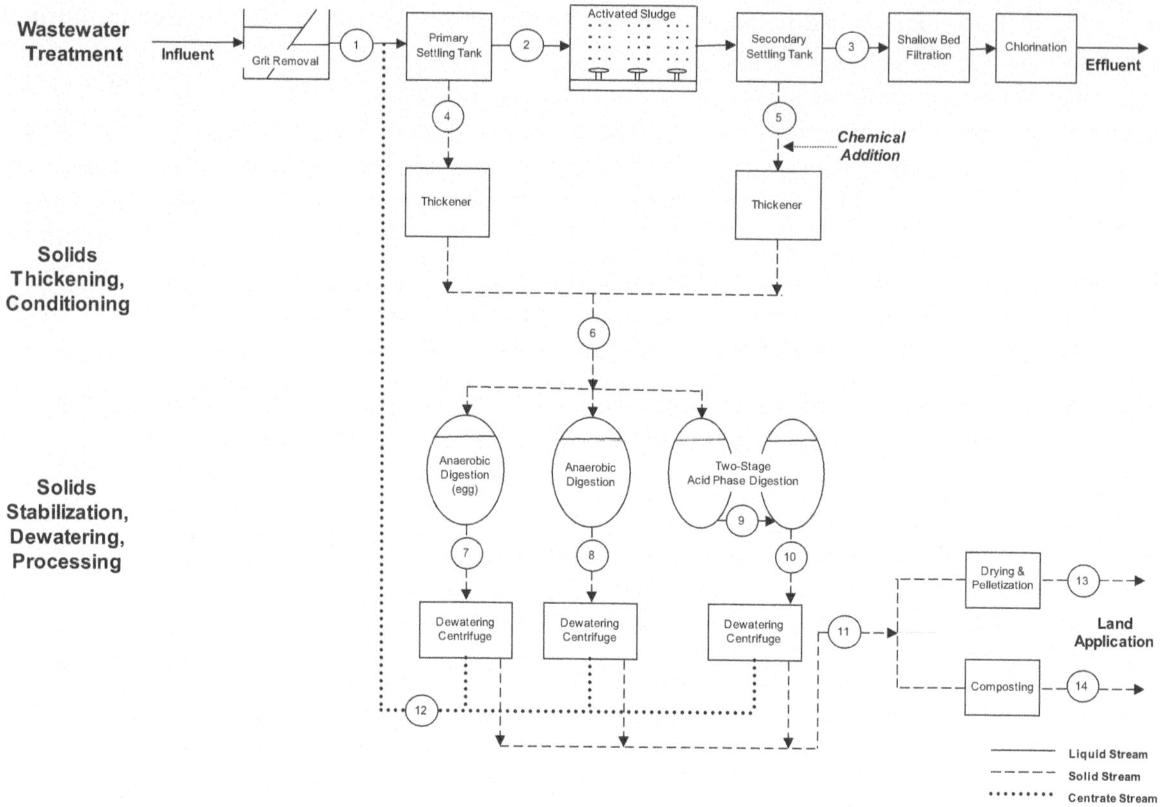

Figure 2-2. Plant B Process Train Schematic.

Table 2-2. Plant B Sampling Points and Frequency.

Figure ID	Location	Sampling Frequency
1	Primary Influent	2x
2	Primary Effluent	2x
3	Secondary Effluent	2x
4	Primary Unthickened Sludge	2x
5	Secondary Unthickened Sludge	2x
6	Thickened Sludge (Combined Primary & Secondary)	3x
7	Conventional Digested Sludge	1x
8	Anaerobically Digested Sludge	5x
9	Acid Phase Digested Sludge	3x
10	Methane Phase Digested Sludge	3x
11	Dewatered Sludge	5x
12	Centrate Recycle Stream from Dewatering Process	5x
13	Tertiary Pelletized Sludge	3x
14	Composted Sludge	1x

2.1.3 Plant C

Plant C treats an average flow of 370 MGD (1,400,600 m^3/d). The existing wastewater treatment process at Plant C consists of preliminary treatment, secondary treatment, nitrification/denitrification, effluent filtration, chlorination/dechlorination and post aeration. Plant C produces an annual average of 135,050 dry tons (122,515,299 kg) per year of biosolids.

Biosolids handling processes include primary sludge screening and degritting, gravity thickening of primary sludge, dissolved air flotation thickening of biological sludge, and centrifuge dewatering of the combined thickened streams. Following dewatering, the sludge is stabilized using lime (15-25% of dry weight) and conveyed to the biosolids storage prior to disposal off-site through land application. The lime stabilized sludge is land applied for agricultural uses, including tree farming. Plant C was selected for this study due to its use of lime stabilization prior to land application.

A schematic and list of the sample points for Plant C are provided in Figure 2-3 and Table 2-3, respectively.

Samples were collected from sample points 1-6 at Plant C on December 6, 2005. Sample points 1-3 were not included in the scope of work for this proposal but were added during this sample trip. Samples were collected from sample points 4-6 on July 17, 2006.

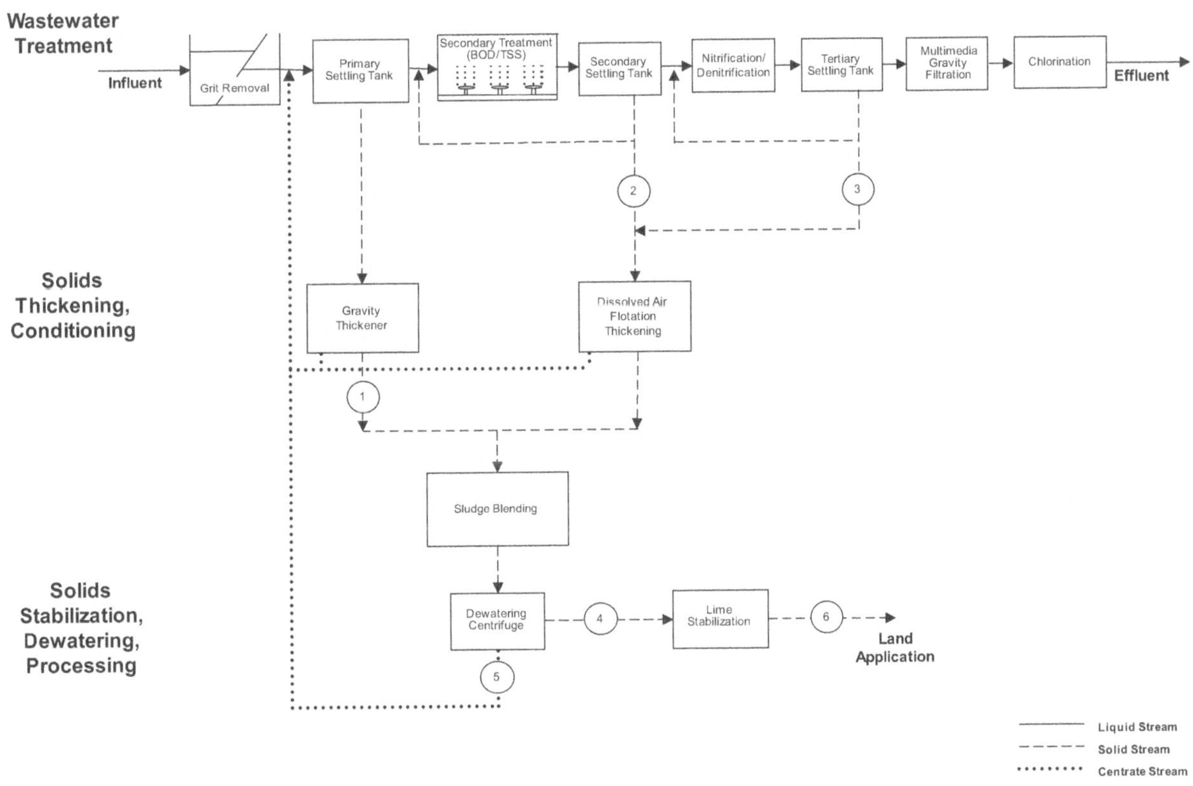

Figure 2-3. Plant C Process Train Schematic.

Table 2-3. Plant C Sampling Points and Frequency.

Figure ID	Location	Sampling Frequency
1	Thickened Primary Waste Sludge	1x
2	Secondary Waste Sludge	1x
3	Nitrification/Denitrification Waste Sludge	1x
4	Dewatered Sludge (Combined Primary & Secondary)	2x
5	Centrate Recycle Stream from Dewatering Process	2x
6	Lime Stabilized Sludge	2x

2.1.4 Plant D

Plant D treats an average flow of 360 MGD (1,362,740 m^3/d). The primary solids production of Plant D is typically about 2.17 MGD (8,214 m^3/d) with a water content of slightly more than 96%. The thickened waste activated sludge (WAS) flow rate is 0.93 MGD (3,520 m^3/d) with a water content of about 96.5%. Plant D was selected to participate in this research program because of the thermophilic anaerobic digestion process operated at that facility. Solids are stabilized by thermophilic anaerobic digestion yielding an average solids flow rate of about 3.1 MGD (11,735 m^3/d), which after dewatering, produces approximately 800 wet tons (725,748 kg) per day of solids for disposal.

Plant D utilizes high-solids, high-capacity centrifuges for dewatering digested biosolids that are capable of processing 600-1000 gallons per minute (gpm) (0.038 - 0.063 m^3/s) of anaerobically digested wastewater sludge while producing a wet cake product in excess of 30% solids. The wet cake product is delivered to storage silos. Stabilized and dewatered biosolids are used as soil nutrients to non-food crops.

A schematic and listing of the sample points for Plant D are provided in Figure 2-4 and Table 2-4, respectively.

Samples were collected from all sample points (1-9) at Plant D on June 19, 2006 and December 5, 2006. A subset of samples was collected on March 16, 2006 (4, 6, 8, and 9) and September 14, 2007 (4, 6, 7, and 8).

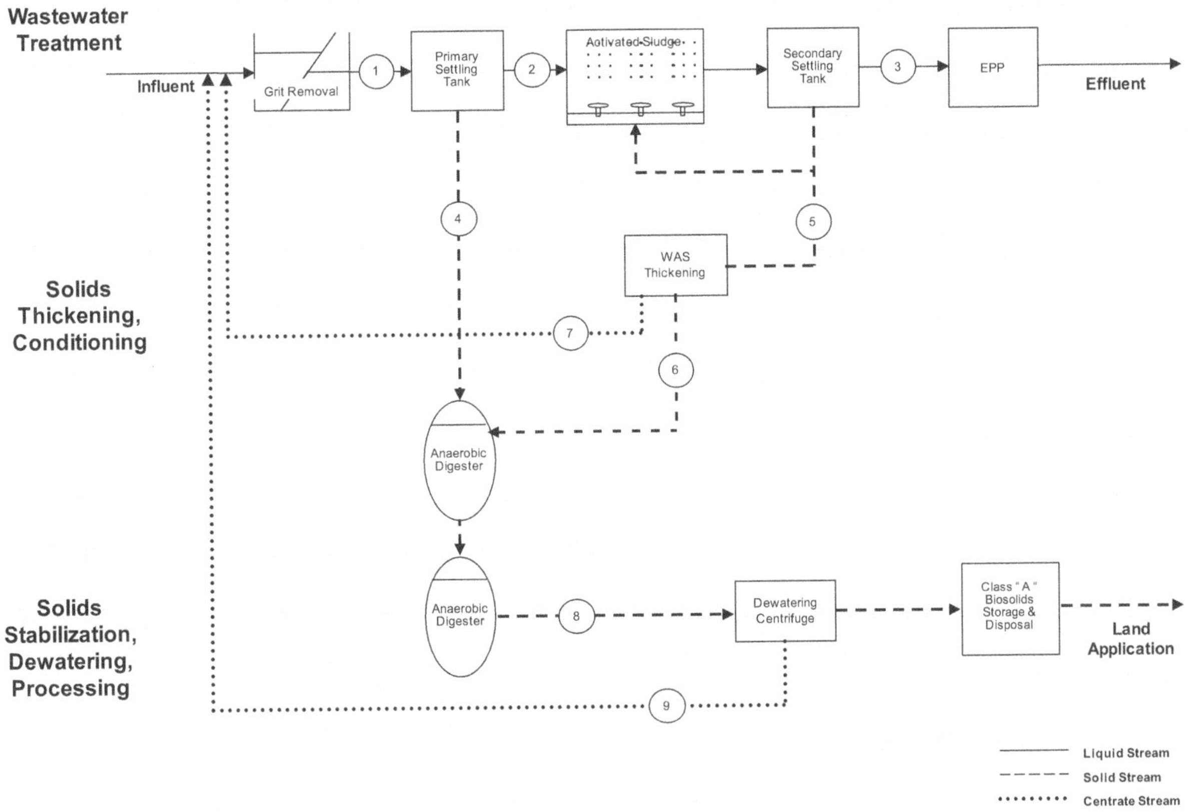

Figure 2-4. Plant D Process Train Schematic.

Table 2-4. Plant D Sampling Points and Frequency.

Figure ID	Location	Sampling Frequency
1	Primary Influent	2x
2	Primary Effluent	2x
3	Secondary Effluent	2x
4	Primary Sludge (Unthickened)	4x
5	Waste Activated Sludge (Unthickened WAS)	2x
6	Thickened Waste Activated Sludge (TWAS)	4x
7	TWAS Centrate	3x
8	Digested Sludge	4x
9	Centrate Recycle Stream from Dewatering Process	3x

2.2 Sample Collection, Preparation and Storage

Both liquid and solid samples were collected in this study. Liquid samples were collected as 24-hour composite samples using automated samplers, for which the frequency of sampling and number of sample locations varied by plant and by sampling schedule. Collection of the 24-hour composite samples from the solids sample points was not possible given sample characteristics, discontinuous flows, and the solids retention time (SRT) of unit processes as well as limited time allocated for sample collection trips (four days maximum per sample trip). As such, a grab sample was collected for each solids sample. Because the solids retention time is typically longer than that of the liquids, solids grab samples should contain a wider averaged time interval of processes (suggested by the observed uniformity of samples from each site throughout the study).

2.2.1 Sampling Equipment

Both liquid and solid samples were collected and handled using teflon, stainless steel, or glass equipment cleaned according to USGS trace-organic protocols (USGS Field Manual, 2009). Liquid samples collected at primary influent, primary effluent, and secondary effluent sites were generally collected using autosamplers with glass bottles cleaned according to USGS trace-organic methods (Shelton, 1994). Unique plant and sample location identification numbers were established for each plant and used consistently throughout the study.

2.2.2 Sample Collection

Liquid and solid samples were collected using standard USGS trace-organic methods (USGS Field Manual, 2009), with some adjustments made based on circumstances specific to WWTPs. Liquid samples were generally collected using autosamplers, with individual autosampler aliquots collected in glass jars that were flow-weight composited in stainless steel or Teflon-lined vessels, previously cleaned according to trace-organic methods, and transferred to 1-L baked glass amber bottles for shipment to the laboratory. Filtration (required for some analytical methods) through a pre-ashed, 0.7 μm nominal pore size glass fiber filter, was performed at the analyzing laboratory (USGS Field Manual, 2009). Similar trace-organic data collection procedures were used for solids samples, which were collected into 500 mL glass amber wide-mouth jars. At some sampling points, disinfection by chlorination or chloramination was employed. Samples collected at these points were preserved with 100 mg ascorbic acid to scavenge residual chlorine and prevent degradation of the compounds of interest. Once collected, samples were stored on ice, shipped by overnight express, on the same day as collection, to the analytical laboratory and refrigerated at 4°C until filtered; samples also were chilled after filtering.

Figure 2-5 shows an overview of the sample collection and analytical procedure for the chemical analyses employed in this task. Figure 2-6 shows an overview of analytical procedures for the bioassay analyses employed on the project.

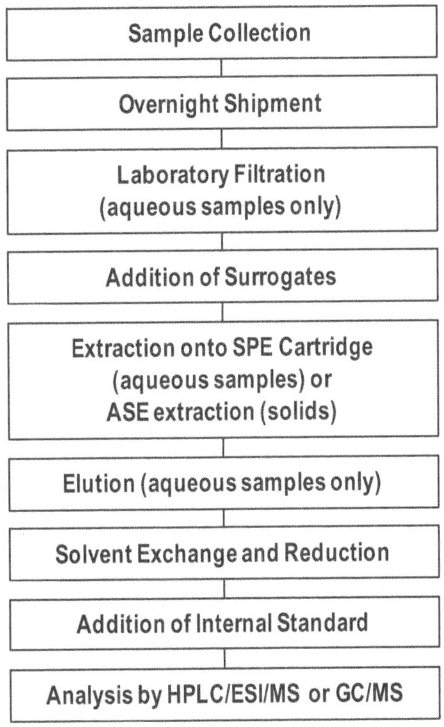

Figure 2-5. Overview of Sample Collection and Analytical Procedure for the Chemical Analyses Employed in this Study.

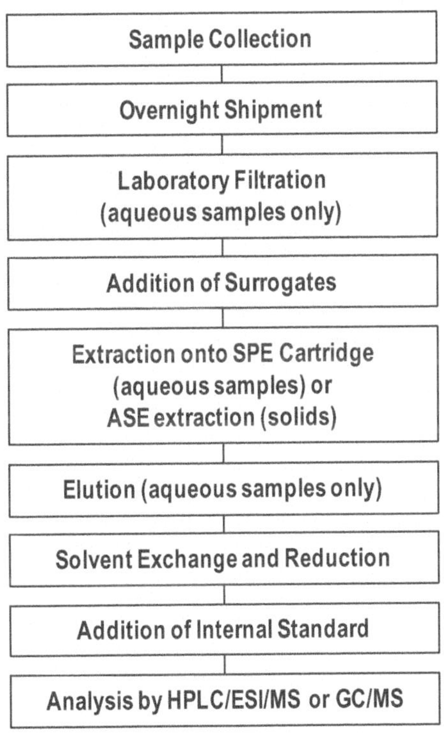

Figure 2-6. Overview of Sample Collection and Analytical Procedure for the Bioassay Analyses Employed in this Study.

2.2.3 Field Quality Assurance Samples

Field quality assurance samples included blanks, sample replicates, and spikes. Laboratory grade organic-free water certified by the USGS was used to prepare blanks, and these blanks were processed in an identical fashion to the environmental samples. Blank samples include equipment blanks and field blanks. For example, blank samples were processed through the entire autosampler assembly to test for potential contamination. Replicate samples also were collected on a regular basis to assess the precision of field sample concentrations. Replicates were collected by splitting environmental samples and sending these as separate samples to participating laboratories.

2.2.4 Sample Handling, Custody and Storage

Each bottle (or jar) was labeled with:

1. Plant name;
2. Unique station identification number;
3. Date and time when the sample was collected;
4. Analytical laboratory;
5. Analytical method to be performed on the sample.

2.2.5 Data Management

All data manipulation (such as concentration calculations) and compilation of analytical results were performed using Microsoft Excel by USGS and UA. All raw and processed data were stored on a central server and archived at least monthly on CD-ROM. This information was disseminated to the rest of the project team via AECOM.

For chemical analysis, because the methods used for bisphenol A, nonylphenol, and nonylphenol ethoxylates (Zaugg et al., 2002) and pharmaceuticals (Cahill et al., 2004, Furlong et al., 2008) were official USGS analytical methods or were in transition to official methods, sample concentration results also were stored in the NWQL laboratory information management system and transmitted to the USGS National Water Information System (NWIS).

2.3 Analytical Methods: Chemical Analysis

Once received at the laboratory, samples were refrigerated (4°C). Prior to extraction and analysis, solid samples were evaluated to determine if the sample required centrifugation to effectively separate low-density solids from co-collected liquids. If this step was required, the separated liquid and solid samples were treated as two separate samples and both processed and analyzed. Field and laboratory blanks and replicate samples were processed using the same methods as field samples. For gas chromatography with tandem mass spectrometry (GC/MS/MS), liquid chromatography with mass spectrometry (LC/MS), and liquid chromatography with tandem mass spectrometry (LC/MS/MS) analyses, samples were typically extracted within 48 hours of receipt. Some exceptions occurred, particularly for the centrate samples, which required additional liquid/solids separation, typically refrigerated centrifugation, in order to produce sufficient extractable sample.

Three separate methods were used to characterize the TOrC compositions of the liquid and solid samples. These methods are referred to herein as the hormone, anthropogenic wastewater indicator, and pharmaceutical methods. The methods are described in detail in the data report published on the USGS website (http://pubs.er.usgs.gov/; Furlong et al., 2010).

2.3.1 Analytical Methods and Reporting Levels

Basic data from chemical analyses was critical to assessing the efficacy of different wastewater treatment processes, thus stringent quality assurance and control procedures were necessary. It is important to note that the compounds investigated are not regulated, and have no environmental concentration criteria. The ambient concentrations for many of these compounds are typically expected to be in the range of 1 part per billion (ppb), and some compounds are more typically found in concentrations in the 10s to 100s of parts per trillion (ppt) in wastewater, or even lower for natural and synthetic hormones. Due to the number of compounds in these methods, mean recoveries in spike samples greater than 60% and recovery variabilities (relative standard deviations) less than 25% were considered acceptable. However, some compounds still failed these criteria because of interferences resulting from the inherently complex sample matrix in wastewater derived solids and liquids, were subject to unstable instrument response, and/or were quantified based on a standard that is only available as a technical mixture. Because of these limitations, these chemicals were reported as estimated concentrations.

A complete list of compounds analyzed in this study, including common name/use, the Chemical Abstracts Service Registration Number (CASRN), and the matrix and USGS method code, is provided in Table 2-5.

The term "Lowest Reporting Level" is used in Table 2-5 and throughout this report. The lowest reporting level is based upon the method detection limit, which is determined statistically according to U.S. EPA methodology (U.S. EPA, 2005) for established methods. For developmental methods (i.e. hormones) it estimated from real samples as 10 times peak-to-peak signal to noise in real samples. In many cases, the reporting limits for an individual sample were adjusted to reflect the inability to process a complete standard volume of sample due to filter plugging, etc., or were adjusted to reflect the presence of co-extracted interferences that could not be removed and which precluded accurate identification during mass spectral analysis.

Table 2-5. Complete List of Compounds Analyzed in this Study (Current USGS Analytical Capabilities).

Compound*	Use	CASRN	Lowest reporting level	Note	Matrix	Schedule or lab code
Wastewater Method (Water)						
1,4-dichlorobenzene	deodorizer	106-46-7	0.5 ug/L	a	water	1433
1-methylnaphthalene	PAH	90-12-0	0.5 ug/L		water	1433
2,6-dimethylnaphthalene	PAH	581-42-0	0.5 ug/L		water	1433
2-methylnaphthalene	PAH	91-57-6	0.5 ug/L		water	1433
3-beta-coprostanol	fecal steroid	360-68-9	2.0 ug/L		water	1433
3-methyl-1(H)-indole (Skatole)	fragrance	83-34-1	1.0 ug/L		water	1433
3-tert-butyl-4-hydroxy anisole (BHA)	antioxidant	25013-16-5	5.0 ug/L	a	water	1433
4-cumylphenol	nonionic detergent metabolite	599-64-4	1.0 ug/L		water	1433
4-n-octylphenol	nonionic detergent metabolite	1806-26-4	1.0 ug/L		water	1433
4-tert-octylphenol	nonionic detergent metabolite	140-66-9	1.0 ug/L		water	1433
5-methyl-1H-benzotriazole	antiocorrosive	136-85-6	2.0 ug/L		water	1433
acetophenone	fragrance	98-86-2	0.5 ug/L		water	1433
acetyl hexamethyl tetrahydronaphthalene (AHTN)	fragrance	21145-77-7	0.5 ug/L		water	1433
anthracene	PAH	120-12-7	0.5 ug/L		water	1433
anthraquinone	pesticide	84-65-1	0.5 ug/L		water	1433
benzo[a]pyrene	PAH	50-32-8	0.5 ug/L		water	1433
benzophenone	plasticizer	119-61-9	0.5 ug/L		water	1433
beta-sitosterol	plant steroid	83-46-5	2.0 ug/L		water	1433
beta-stigmastanol	plant steroid	19466-47-8	2.0 ug/L		water	1433
bisphenol A	plasticizer	80-05-7	1.0 ug/L		water	1433
bromacil	herbicide	314-40-9	0.5 ug/L		water	1433
bromoform	disinfectant	75-25-2	0.5 ug/L	a	water	1433
caffeine	stimulant	58-08-2	0.5 ug/L		water	1433
camphor	flavorant	76-22-2	0.5 ug/L		water	1433
carbaryl	insecticide	63-25-2	1.0 ug/L	b	water	1433
carbazole	PAH	86-74-8	0.5 ug/L		water	1433
chlorpyrifos	insecticide	2921-88-2	0.5 ug/L		water	1433
cholesterol	plant/animal steroid	57-88-5	2.0 ug/L		water	1433
cotinine	nicotine metabolite	486-56-6	1.0 ug/L		water	1433
decafluorobiphenyl	polymer	-	pct		water	1433
d-limonene	fungicide	5989-27-5	0.5 ug/L	a	water	1433
fluoranthene	PAH	206-44-0	0.5 ug/L		water	1433
hexadydrohexamethylcyclopentabenzopyran (HHCB)	fragrance	1222-05-5	0.5 ug/L		water	1433
indole	pesticide inert ingredient	120-72-9	0.5 ug/L		water	1433
isoborneol	fragrance	124-76-5	0.5 ug/L		water	1433
isophorone	solvent	78-59-1	0.5 ug/L		water	1433
isoquinoline	fragrance	119-65-3	0.5 ug/L		water	1433
menthol	flavorant	89-78-1	0.5 ug/L		water	1433
metalaxyl	pesticide	57837-19-1	0.5 ug/L		water	1433
methyl salicylate	liniment	119-36-8	0.5 ug/L		water	1433
N,N-diethyl-meta-toluamide (DEET)	insect repellant	134-62-3	0.5 ug/L		water	1433
nonylphenol, diethoxy-(total) (NPEO2)	nonionic detergent metabolite	26027-38-2	5.0 ug/L	c	water	1433
octylphenol, diethoxy- (OPEO2)	nonionic detergent metabolite	26636-32-8	1.0 ug/L	c	water	1433

Table 2-5. Complete List of Compounds Analyzed in this Study (Current USGS Analytical Capabilities) (continued).

Compound*	Use	CASRN	Lowest reporting level	Note	Matrix	Schedule or lab code
octylphenol, monoethoxy- (OPEO1)	nonionic detergent metabolite	26636-32-8	1.0 ug/L	c	water	1433
para-nonylphenol (total)	nonionic detergent metabolite	84852-15-3	5.0 ug/L	c	water	1433
p-cresol	antioxidant	106-44-5	1.0 ug/L		water	1433
pentachlorophenol	pesticide	87-86-5	2.0 ug/L	b	water	1433
phenanthrene	PAH	85-01-8	0.5 ug/L		water	1433
phenol	disinfectant	108-95-2	0.5 ug/L		water	1433
pyrene	PAH	129-00-0	0.5 ug/L		water	1433
tetrachloroethylene	solvent, degreaser	127-18-4	0.5 ug/L		water	1433
tri(2-butoxyethyl)phosphate	fire retardant	78-51-3	0.5 ug/L		water	1433
tri(2-chloroethyl)phosphate	fire retardant	115-96-8	0.5 ug/L		water	1433
tri(dichlorisopropyl)phosphate	fire retardant	13674-87-8	0.5 ug/L		water	1433
tributyl phosphate	fire retardant	126-73-8	0.5 ug/L		water	1433
triclosan	antimicrobial disinfectant	3380-34-5	1.0 ug/L		water	1433
triethyl citrate (ethyl citrate)	plasticizer	77-93-0	0.5 ug/L		water	1433
triphenyl phosphate	plasticizer	115-86-6	0.5 ug/L		water	1433
Wastewater Method (Sediment)						
1,4-dichlorobenzene	deodorizer	106-46-7	50 ug/kg		sediment	8050
1-methylnaphthalene	PAH	90-12-0	50 ug/kg		sediment	8050
2,6-dimethylnaphthalene	PAH	581-42-0	50 ug/kg		sediment	8050
2-methylnaphthalene	PAH	91-57-6	50 ug/kg		sediment	8050
3,4-dichlorophenyl isocyanate	plastic additive	102-36-3	100 ug/kg	a	sediment	8050
3-beta-coprostanol	fecal steroid	360-68-9	500 ug/kg		sediment	8050
3-methyl-1(H)-indole (Skatole)	fragrance	83-34-1	50 ug/kg		sediment	8050
3-tert-butyl-4-hydroxy anisole (BHA)	antioxidant	25013-16-5	100 ug/kg	a	sediment	8050
4-cumylphenol	nonionic detergent metabolite	599-64-4	50 ug/kg		sediment	8050
4-n-octylphenol	nonionic detergent metabolite	1806-26-4	50 ug/kg		sediment	8050
4-tert-octylphenol	nonionic detergent metabolite	140-66-9	50 ug/kg		sediment	8050
acetophenone	fragrance	98-86-2	100 ug/kg	a	sediment	8050
acetyl hexamethyl tetrahydro-naphthalene (AHTN)	fragrance	21145-77-7	50 ug/kg		sediment	8050
anthracene	PAH	120-12-7	50 ug/kg		sediment	8050
anthraquinone	pesticide	84-65-1	50 ug/kg		sediment	8050
atrazine	herbicide	1912-24-9	100 ug/kg		sediment	8050
benzo[a]pyrene	PAH	50-32-8	50 ug/kg		sediment	8050
benzophenone	plasticizer	119-61-9	50 ug/kg		sediment	8050
beta-sitosterol	plant steroid	83-46-5	500 ug/kg		sediment	8050
beta-stigmastanol	plant steroid	19466-47-8	500 ug/kg		sediment	8050
bisphenol A	plasticizer	80-05-7	50 ug/kg	a	sediment	8050
bromacil	herbicide	314-40-9	500 ug/kg	a	sediment	8050
camphor	flavorant	76-22-2	50 ug/kg		sediment	8050
carbazole	PAH	86-74-8	50 ug/kg		sediment	8050
chlorpyrifos	insecticide	2921-88-2	50 ug/kg		sediment	8050
cholesterol	plant/animal steroid	57-88-5	250 ug/kg		sediment	8050
diethyl phthalate	plastic additive	84-66-2	100 ug/kg		sediment	8050
diethylhexyl phthalate	plastic additive	117-81-7	250 ug/kg		sediment	8050
d-limonene	fungicide	5989-27-5	50 ug/kg	a	sediment	8050
fluoranthene	PAH	206-44-0	50 ug/kg		sediment	8050

Table 2-5. Complete List of Compounds Analyzed in this Study (Current USGS Analytical Capabilities) (continued).

Compound*	Use	CASRN	Lowest reporting level	Note	Matrix	Schedule or lab code
hexahydrohexamethylcyclopentabenzopyran (HHCB)	fragrance	1222-05-5	50 ug/kg		sediment	8050
indole	pesticide inert ingredient	120-72-9	50 ug/kg		sediment	8050
isoborneol	fragrance	124-76-5	50 ug/kg		sediment	8050
isophorone	solvent	78-59-1	50 ug/kg	a	sediment	8050
isopropylbenzene (cumene)	solvent	98-82-8	100 ug/kg	a	sediment	8050
isoquinoline	fragrance	119-65-3	100 ug/kg	a	sediment	8050
menthol	flavorant	89-78-1	50 ug/kg		sediment	8050
metalaxyl	pesticide	57837-19-1	50 ug/kg	a	sediment	8050
methyl salicylate	liniment	119-36-8	50 ug/kg	a	sediment	8050
metolachlor	herbicide	51218-45-2	50 ug/kg		sediment	8050
N,N-diethyl-meta-toluamide (DEET)	insect repellant	134-62-3	50 ug/kg	a	sediment	8050
naphthalene	PAH	91-20-3	50 ug/kg		sediment	8050
nonylphenol, diethoxy-(total) (NPEO2)	nonionic detergent metabolite	26027-38-2	1000 ug/kg	d	sediment	8050
nonylphenol, monoethoxy-(total) (NPEO1)	nonionic detergent metabolite		500 ug/kg	d	sediment	8050
octylphenol, diethoxy- (OPEO2)	nonionic detergent metabolite	26636-32-8	50 ug/kg	d	sediment	8050
octylphenol, monoethoxy- (OPEO1)	nonionic detergent metabolite	26636-32-8	250 ug/kg	d	sediment	8050
para-nonylphenol (total)	nonionic detergent metabolite	84852-15-3	250 ug/kg	d	sediment	8050
para-cresol	antioxidant	106-44-5	500 ug/kg		sediment	8050
pentachlorophenol	pesticide	87-86-5	500 ug/kg	a	sediment	8050
phenanthrene	PAH	85-01-8	50 ug/kg		sediment	8050
phenol	disinfectant	108-95-2	50 ug/kg	a	sediment	8050
pyrene	PAH	129-00-0	50 ug/kg		sediment	8050
2,2',4,4'-tetrabromodiphenyl ether	fire retardant	40088-47-9	50 ug/kg		sediment	8050
tri(2-butoxyethyl)phosphate	fire retardant	78-51-3	100 ug/kg		sediment	8050
tri(2-chloroethyl)phosphate	fire retardant	115-96-8	100 ug/kg		sediment	8050
tri(dichlorisopropyl)phosphate	fire retardant	13674-87-8	100 ug/kg		sediment	8050
tributyl phosphate	fire retardant	126-73-8	50 ug/kg		sediment	8050
triclosan	antimicrobial disinfectant	3380-34-5	50 ug/kg		sediment	8050
triphenyl phosphate	plasticizer	115-86-6	50 ug/kg	a	sediment	8050
Human Pharmaceuticals Method (Water)						
1,7-dimethylxanthine	caffeine metabolite	611-59-6	0.144 ug/L	e	water	LC 9003
codeine	analgesic	76-57-3	0.015 ug/L	e	water	LC 9003
caffeine	stimulant	58-08-2	0.016 ug/L	e	water	LC 9003
thiabendazole	fungicide		0.011 ug/L	e	water	LC 9003
albuterol (Salbutamol)	antiasthmatic	18559-94-9	0.023 ug/L	e	water	LC 9003
acetaminophen	antipyretic	103-90-2	0.036 ug/L	e	water	LC 9003
cotinine	nicotine metabolite	486-56-6	0.014 ug/L	e	water	LC 9003
dehydronifedipine	nifedipine metabolite	67035-22-7	0.015 ug/L	e	water	LC 9003
carbamazapine	anticonvulsant		0.011 ug/L	e	water	LC 9003
trimethoprim	antibiotic	738-70-5	0.013 ug/L	e	water	LC 9003
warfarin	anticoagulant	81-81-2	0.012 ug/L	e	water	LC 9003
diphenhydramine	antihistamine		0.015 ug/L	f	water	LC 9003
sulfamethoxazole	antibiotic	723-46-6	0.064 ug/L	f	water	LC 9003
diltiazem	antihypertensive	42399-41-7	0.016 ug/L	f	water	LC 9003

Table 2-5. Complete List of Compounds Analyzed in this Study (Current USGS Analytical Capabilities) (continued).

Compound*	Use	CASRN	Lowest reporting level	Note	Matrix	Schedule or lab code
ibuprofen	antiinflammatory	15687-27-1	0.042 ug/L	f	water	LC 9003
ranitidine	antacid	66357-35-5	0.013 ug/L	f	water	LC 9003
cimetidine	antacid	51481-61-9	0.012 ug/L	f	water	LC 9003
fluoxetine	antidepressant	54910-89-3	0.014 ug/L	f	water	LC 9003
gemfibrozil	antihyperlipidemic	25812-30-0	0.013 ug/L	f	water	LC 9003
erythromycin	antibiotic	114-07-8	0.009 ug/L	g	water	LC 9003
azithromycin	antibotic		0.004 ug/L	g	water	LC 9003
miconazole	antifungal		0.018 ug/L	g	water	LC 9003
metformin	antidiabetic	657-24-9	N/D	g	water	LC 9003
Human Pharmaceuticals Method (Sediment)						
1,7-dimethylxanthine	caffeine metabolite	1611-59-6			sediment	LC 9008
codeine	analgesic	76-57-3			sediment	LC 9008
caffeine	stimulant	58-08-2			sediment	LC 9008
thiabendazole	fungicide				sediment	LC 9008
albuterol (Salbutamol)	antiasthmatic	18559-94-9			sediment	LC 9008
acetaminophen	antipyretic	103-90-2			sediment	LC 9008
cotinine	nicotine metabolite	485-56-6			sediment	LC 9008
dehydronifedipine	nifedipine metabolite	67035-22-7			sediment	LC 9008
carbamazapine	anticonvulsant				sediment	LC 9008
trimethoprim	antibiotic	738-70-5			sediment	LC 9008
warfarin	anticoagulant	81-81-2			sediment	LC 9008
diphenhydramine	antihistamine				sediment	LC 9008
sulfamethoxazole	antibiotic	723-46-6			sediment	LC 9008
diltiazem	antihypertensive	42399-41-7			sediment	LC 9008
ranitidine	antacid	66357-35-5			sediment	LC 9008
cimetidine	antacid	51481-61-9			sediment	LC 9008
fluoxetine	antidepressant	54910-89-3			sediment	LC 9008
erythromycin	antibiotic	114-07-8			sediment	LC 9008
miconazole	antifungal				sediment	LC 9008
azithromycin	antibiotic				sediment	LC 9008
ibuprofen	antiinflammatory	15687-27-1			sediment	LC 9008
gemfibrozil	antihyperlipidemic	25812-30-0			sediment	LC 9008
furosemide	diuretic				sediment	LC 9008
metformin	antidiabetic	657-24-9			sediment	LC 9008
Hormone Method (Water)						
cis-androsterone	urinary steroid	53-41-8			water	SH4434
4-Androsten-3,17-dione	natural androgen	63-05-8			water	SH4434
cholesterol	plant/animal steroid	57-88-5			water	SH4434
3-beta-coprostanol	animal fecal steroid	360-68-9			water	SH4434
Diethylstilbestrol	synthetic estrogen	56-53-1			water	SH4434
Epitestosterone	natural androgen	481-30-1			water	SH4434
equilenin	hormone replacement	517-09-9			water	SH4434
equilin	hormone replacement	474-86-2			water	SH4434
17-alpha-estradiol	reproductive hormone	57-91-0			water	SH4434
17-beta-estradiol	reproductive hormone	50-28-2			water	SH4434
estriol	reproductive hormone	50-27-1			water	SH4434
estrone	reproductive hormone	53-16-7			water	SH4434

Table 2-5. Complete List of Compounds Analyzed in this Study (Current USGS Analytical Capabilities) (continued).

Compound*	Use	CASRN	Lowest reporting level	Note	Matrix	Schedule or lab code
17-alpha-ethynylestradiol	ovulation inhibitor	57-63-6			water	SH4434
11-ketotestosterone	natural androgen	564-35-2			water	SH4434
mestranol	ovulation inhibitor	72-33-3			water	SH4434
19-norethisterone	ovulation inhibitor	68-22-4			water	SH4434
progesterone	reproductive hormone	57-83-0			water	SH4434
stanolone	natural androgen	521-18-6			water	SH4434
testosterone	reproductive hormone	58-22-0			water	SH4434
Hormone Method (Sediment)						
cis-androsterone	urinary steroid	53-41-8			sediment	SH6434
4-Androsten-3,17-dione	natural androgen	63-05-8			sediment	SH6434
cholesterol	plant/animal steroid	57-88-5			sediment	SH6434
3-beta-coprostanol	animal fecal steroid	360-68-9			sediment	SH6434
Diethylstilbestrol	synthetic estrogen	56-53-1			sediment	SH6434
Epitestosterone	natural androgen	481-30-1			sediment	SH6434
equilenin	hormone replacement	517-09-9			sediment	SH6434
equilin	hormone replacement	474-86-2			sediment	SH6434
17-alpha-estradiol	reproductive hormone	57-91-0			sediment	SH6434
17-beta-estradiol	reproductive hormone	50-28-2			sediment	SH6434
estriol	reproductive hormone	50-27-1			sediment	SH6434
estrone	reproductive hormone	53-16-7			sediment	SH6434
17-alpha-ethynylestradiol	ovulation inhibitor	57-63-6			sediment	SH6434
11-ketotestosterone	natural androgen	564-35-2			sediment	SH6434
mestranol	ovulation inhibitor	72-33-3			sediment	SH6434
19-norethisterone	ovulation inhibitor	68-22-4			sediment	SH6434
progesterone	reproductive hormone	57-83-0			sediment	SH6434
stanolone	natural androgen	521-18-6			sediment	SH6434
testosterone	reproductive hormone	58-22-0			sediment	SH6434

Notes:

*	known or suspected hormonally active agents are in bold.
a	concentration is estimated because recovery is less than 60% or precision is greater than 25% RSD.
b	concentration is always estimated because of unstable instrument response
c	concentration is always estimated because the reference standard is from a technical mixture
d	concentration is estimated because the reference standard is from a technical mixture
e	recovery > 60%
f	recovery 30-60%
g	recovery < 30%

2.3.1.1 Anthropogenic Wastewater Indicators (AWIs)

The term AWI is used herein to describe a wide array of TOrCs including, personal care products, detergent metabolites, flame retardants, and pesticides. The methods used for these TOrCs, which includes the EDCs bisphenol A, nonylphenol, nonylphenol ethoxylates and other AWIs in filtered liquid and solid samples are described in Zaugg et al. (2002) Burkhardt et al. (2006), respectively. Briefly, for liquids, the analytes in a 1-L filtered sample were extracted by passing the sample through a cartridge containing 0.5 g of a modified polystyrene-divinylbenzene solid-phase extraction (SPE) phase (Oasis HLB; Waters Corp., Milford, MA) at a flow rate of between 25 and 50 mL/minute. The cartridges were then thoroughly dried under nitrogen and the TOrCs eluted with 15 mL of dichloromethane:diethyl ether (4:1). The extracts were reduced to a few mL under a gentle steam of nitrogen, a suite of internal standards added, and the extract reduced to a final volume of 0.4 mL before analysis by full-scan electron-impact ionization gas chromatography with mass spectrometry (GC/MS).

Up to 10 g (wet weight; more typically 1 g or less) of a solids sample was extracted using accelerated solvent extraction (Burkhardt et al., 2006), using a two-step extraction program. Extraction was first performed at 120°C with water/isopropanol (50:50, v/v) to obtain the major portion of polar and heat susceptible compounds. The same cell then was extracted with water/IPA (20:80, v/v) at 200°C. All extractions were performed at 13,800 kPa and each extraction consisted of three 10-minute static extraction cycles at each temperature. Each 40-mL extract was diluted with 100 mL of a phosphate buffer (pH 7), and then the diluted 200°C extract is passed through a 0.5 g Oasis SPE cartridge, followed by the diluted 120°C extract. Four grams of anyhydrous sodium sulfate was added to a 1-g fluorisil SPE cartridge and this cartridge was attached below the Oasis SPE cartridge. The tandem SPE cartridge set was then eluted with three 10-ml aliquots of dichloromethane: diethyl ether (4:1), the extracts concentrated, amended with an aliquot of the internal standard solution, reduced to a final volume of 0.4 mL, and the extracts analyzed by full-scan electron-impact ionization GC/MS.

2.3.1.2 Pharmaceuticals

The analysis of human-health pharmaceuticals from liquids is described in Cahill et al. (2004). The pharmaceuticals in a 1-L filtered sample were extracted by passing the sample through a cartridge containing 0.5 g of a modified polystyrene-divinylbenzene SPE phase (Oasis HLB; Waters Corp., Milford, Mass.) at a flow rate of 15 mL/min. After extraction, the SPE cartridge was dried with air, and the adsorbed pharmaceuticals were eluted from the dried cartridge by using two sequential elutions of 1) 6 mL methanol followed by 2) 4 mL of methanol, acidified with trifluoroacetic acid (0.1%). The resulting sample extracts were reduced under nitrogen to near dryness (approximately 0.1 mL), and then reconstituted to a volume of 1.0 ± 0.1 mL with the initial high-performance liquid chromatography (HPLC) eluent, aqueous ammonium formate/formic acid buffer (10 mmol, pH 3.7). The pharmaceuticals were chromatographically separated by HPLC using a reverse-phase octadecylsilane HPLC column and an aqueous formate buffer:acetonitrile gradient. The HPLC was coupled to the quadrupole MS by an electrospray ionization interface, and the separated pharmaceuticals were detected, identified, and quantified using electrospray ionization operated in the positive ion mode using selected-ion monitoring (SIM) to improve specificity and reduce chemical noise.

Pharmaceuticals in sediment were determined by the method described in Kinney et al. (2006a, b) for the analysis of soils and biosolids. Briefly, an aliquot of wet solids, equivalent to no more than 10 grams of dry solids, was extracted by using accelerated solvent extraction, which minimized degradation of these polar, labile compounds. Three sequential extractions were carried out using 70% acetonitrile/30% water at a temperature of 130°C and a pressure of 10.34 x 107 pascals (1,500 per square inch). Typically, the final volume of extract was approximately 20 ml. A 1-ml extract subsample was filtered using a 0.20-mm syringe filter into a HPLC vial, and then the acetonitrile was evaporated under nitrogen. The concentrated aqueous extract volume (~0.3 ml) was increased to 1 ml with 0.050 ml of a 1.59 x 10-4 mM nicotinamide-2,4,5,6-d4 solution, added as an internal standard, and approximately 0.65 ml of a 10-mM aqueous ammonium formate buffer. The sediment extracts were analyzed in a similar manner to the liquid extracts, using the method of Cahill et al. (2004), but for a somewhat different list of pharmaceuticals, reflecting the differing propensities of pharmaceuticals to associate with solids. The presence of pharmaceuticals was confirmed in select sample extracts using an HPLC/MS/MS method analogous to the SIM-HPLC/MS method of Cahill et al. (2004), currently in development at the National Water Quality Laboratory.

2.3.1.3 Steroid Hormones

A suite of 19 steroid hormones, including estrogens and androgens, were isolated from water and solids and analyzed by GC/MS/MS. Liquid sample isolation was based on the procedures outlined in Barber et al. (2005). Solid samples were extracted using accelerated solvent extraction. The extraction and cleanup procedures were by the same technique as the wastewater indicator compounds (Burkhardt et al., 2005) with minor modifications to solvent composition for enhanced recovery of estriol and diethylstilbestrol. Following extraction, steroids were separated from interfering natural organic matter based polarity using florisil cartridges. After cleanup, extracts were derivatized using activated MSTFA. Steroid derivatives were then quantified by isotope dilution based on twelve deuterated surrogate standards.

The addition of the surrogate compounds has greatly enhanced quantitative accuracy in difficult matrices. Surrogates were added prior to extraction (liquid and solid sample) and carried through the entire process. If matrix interference or loss of an analyte through cleanup and derivatization occurred in an individual sample it was evident in low recovery of the surrogate standard and accounted for in final quantitation. This is particularly important for some biosolid samples that were processed before method development was complete because high levels of derivatizable material in the extracts resulted in certain cases where reaction yields were low. With the addition of surrogates, recovery and reproducibility was very good across this wide variety of sample matrices (Table 2-6). Similar enhancements in data quality are evident in solid samples analyzed in conjunction with this project. In 2010, the USGS will complete development and validation of the hormones in solids method, which will be published as an official USGS method.

Recognizing the substantial impact the sample matrix can have upon analyte recoveries from liquid and solid samples is critical to interpreting the results from the chemical methods used in this study. These methods were initially developed for application in surface water, treated effluent, soils, and streambed sediment. The extension of these methods to liquid and solid waste samples collected from earlier stages in the wastewater treatment process may result in quantitative results that can be substantially affected by higher concentrations of organic matter present in these sample types. The effect of sample matrix can be twofold: first by interference in sample extraction, where the matrix plugs the SPE cartridge (liquid), or is amorphous and has very low dry solids content (solids) and reduces the processed sample volume, or by competing for sorptive sites on the SPE cartridge, reducing the recovery of the analytes. The second form of matrix effect is interference during instrumental analysis, where ionized matrix components coelute from the chromatographic separation with the compounds of interest and are present in the mass spectrum, or, in the case of HPLC/MS analysis, where the sample matrix can suppress or enhance ionization of the analytes of interest, potentially adding bias to the results. The use of an isotope-dilution approach for analyzing the hormones by GC/MS/MS reduces, but does not eliminate, the potential for bias introduced by sample matrix.

The primary matrix effect on extraction observed in this study was to reduce the processed sample volume. For example, in the case of solids extracted by accelerated solvent extraction (ASE) for analytes by GC/MS, the mean and median processed dry masses were 0.248 and 0.066 grams, respectively, while the mean and median processed volumes for liquid samples analyzed by SPE and GC/MS were 430 and 244 mL, respectively. These reduced sample volumes resulted in reporting limits that were raised, sometimes substantially, in proportion to the standard volume the method was designed to use (10 grams dry mass for solids, 1,000 mL for liquids). The raised reporting levels vary inversely with the sample volume, which can make comparison between samples difficult. Also it can result in detected concentrations for some compounds that are lower than the reporting levels for other compounds in the same sample or that are lower than the reporting level for the same compound in different samples.

Table 2-6. Summary of Results from Hormone Recovery Experiments in Various Waters (% Recovered).

Sample	Boulder Creek at 95th St.		Plant B		Boulder Primary Effluent		Vail Tertiary Effluent	
Level Spiked (ng/L)	10	100	10	100	10	100	10	100
N	8	8	9	9	9	9	8	7
Analyte								
Diethylstilbestrol	55.6(15.8)	77.2(17.0)	78.3(7.8)	87.2(2.8)	91.6(4.7)	80.8(15.4)	88.4(10.3)	93.4(10.1)
cis-Androsterone	115(19.4)	132.2(9.7)	108.8(10.3)	104.5(4.3)	I[b]	I	115.1(12.4)	108(9.6)
Epitestosterone	87.7(12.8)	94.4(11.4)	104.7(6.4)	105.6(7.6)	139.5(9.9)	93.4(17.3)	106.6(11.7)	108(9.2)
17-alpha-estradiol	83.5(14.8)	99.4(8.0)	106.8(9.9)	110.6(5.3)	104.0(5.8)	84.2(15.5)	103.4(9.8)	115(7.9)
Stanolone	86.4(13.8)	102.2(10.9)	89.1(12.8)	97.0(5.7)	I	104(20.5)	103.8(7.3)	102(8.1)
Androstene-3,17 dione	73.8(14.6)	99.1(13.8)	95.5(12.2)	92.6(5.4)	I	141.4(21.5)	91.5(12.6)	99.9(5.3)
Estrone	80.9(13.7)	106.7(12.4)	100.0(10.8)	100.3(4.4)	97.6(6.8)	77.5(16.8)	100.4(9.2)	108(8.2)
Testosterone	73.5(15.4)	83.9(11.5)	84.8(8.1)	89.5(4.1)	92.6(15.3)	84.5(19.0)	100.5(8.4)	102(10.4)
Equilin	102.3(25.4)	108.3(14.6)	91.0(14.1)	100.3(9.0)	105.4(16.8)	59.8(18.0)	97.9(19.1)	136(48.7)
11-ketotestosterone	52.1(20.8)	58.5(18.0)	93.9(32.0)	109.5(6.4)	76.4(20.7)	55.2(22.7)	104.5(12.3)	104(9.8)
17-beta-estradiol	84.1(12.3)	98.1(9.5)	92.8(9.2)	95.8(4.7)	105.7(3.8)	86.7(15.5)	100(10.1)	105(6.7)
19-norethindrone	76.7(13.5)	91.8(6.7)	84.4(5.7)	86.1(4.3)	96.3(5.7)	82.2(15.5)	92.8(9.3)	95.0(6.3)
Mestranol	84.4(13.2)	96.5(8.0)	93.8(6.8)	96.6(3.2)	104(3.4)	85.5(13.4)	98.1(7.8)	105(6.5)
Equilenin	55.8(16.2)	75.3(25.9)	98.2(10.3)	101.3(9.4)	123(5.3)	97.9(16.2)	83.5(13.5)	68.4(46.6)
17-alpha-EE2	75.5(12)	90.1(7.4)	90.2(7.9)	90.9(3.1)	89.4(3.5)	15.6(16.1)	93.9(11.4)	101(7.3)
Estriol	84.3(15.7)	102.4(8.4)	92.9(9.6)	96.0(2.0)	122(27.2)	91.1(54.3)	98.0(9.6)	104(5.8)
Progesterone	NR[a]	NR	94.8(7.6)	91.3(3.7)	I	I	102.7(11.2)	103(7.7)
Coprostanol	74.3(41.6)	95.2(9.8)	154.4(40.3)	103.5(4.5)	I	I	I	107(12.2)
Cholesterol	96.4(83.4)	93.7(9.1)	151.7(55.3)	100.7(5.6)	I	I	I	87.6(21.2)

Notes: NR = compound was not recovered in this matrix, I = ambient level interference too high to perform recovery calculation. Data presented as Recovery % (relative standard deviation) for N=7-9 replicates for each spiking level. For compound names in italics greater variability occurs due to the lack of an exact isotope-dilution surrogate.

2.3.2 Analytical Methods and Reporting Levels

Extractions were carried out in groups of up to 10 environmental samples, with an additional two laboratory quality control samples in each batch. The first QC category consisted of HPLC grade water (aqueous samples) or ashed Ottowa Sand (solid sample) amended with the performance surrogate, and are referred to as laboratory blanks. The second QC category consisted of HPLC grade water (aqueous samples) or an ashed Ottowa Sand (solid sample) amended with the analytes determined in the method, as well as the performance surrogate, and are referred to as the laboratory matrix spike sample. For every 10 samples, two replicate samples were collected. One was analyzed as a field duplicate, while the second replicate was amended with method analytes and analyzed as a laboratory matrix spike sample. A multipoint internal standard calibration was used for each sample set analyzed. Calibration was monitored through the use of continuing calibration verification (CCV) samples. If the calibration was within ± 20%, analysis of the laboratory QC and environmental samples continued. Instrumental blanks were interspersed between sample sets behind the CCVs to monitor potential carryover between injections. Table 2-7 is an overview of the QC parameters.

Table 2-7. Quality Control Samples.

QC Sample Type	Frequency	Acceptance Criteria	Corrective Action
Performance Surrogate	Every sample	60-120 Percent Recovery	Qualify detections in sample
Laboratory Reagent Blank	One every batch analyzed		Censor/qualify environmental detections that are less than 10 X the blank detection
Laboratory Reagent Spike	One every batch analyzed	60-120 Percent Recovery	Qualify compound-specific results
Intralaboratory Duplicate	1 in 8 collected samples	Relative percent difference = 70-130%	Discuss variability in report
Continuing Calibration Verification	1 every 6 injections	± 20 %	Reanalyze samples that fall outside performance window
Instrument Blank	1 every 6 injections		Perform corrective maintenance

2.4 Analytical Methods: Biological Analysis

Two bioassays were used in this study. The YES bioassay was used for all samples. The YES bioassay is the most widely used yeast-based reporter gene assay (GWRC, 2008). A relatively newer bioassay, KBluc, was used to analyze select solids samples. It is known that different bioassays respond differently to particular estrogenic compounds and in different water matrices. Consequently the utilization of a second bioassay broadened the information gleaned from the sampling effort and provided further cross-comparison between bioassay and single compound methods of quantifying a sample's estrogenic signal.

2.4.1 Sample Preparation

2.4.1.1 Centrifugation and Filtration

Liquid-phase samples (raw influent, primary clarifier effluent, secondary clarifier effluent, effluent from dewatering of thickened sludge, and centrate from the dewatering process after anaerobic digestion), were separated into liquid and solid fractions using a Beckman centrifuge with a JA-10 rotor (20 minutes, RCF = 17,000). Liquid portions were decanted and filtered using 3.1 μm and 0.7 μm Pall glass fiber filters. The liquid and solid fraction of each sample was analyzed for estrogenic activity.

Solid-phase samples (sludges/biosolids) were separated via centrifugation; liquid centrates were not tested because it was assumed most hydrophobic compounds would be found in the particulate fractions.

2.4.1.2 Microwave Assisted-Extraction

After centrifugation/filtration, all solid samples (biosolids/sludges) were extracted in methanol using a microwave-accelerated extraction (MAE) procedure. About 1-g (dry weight) of solid was suspended in 20 mL methanol and extracted at constant pressure (20 psig for 30 min.) using a CEM-MDS 2100 Microwave Digestion System. Reactor contents were cooled for 45 minutes inside the microwave unit before liquids were decanted into muffled glass vials. Methanol extracts were evaporated to 1 mL under nitrogen gas.

2.4.1.3 Separation on C-18 Resin

Solid-phase microwave extracts were diluted to 1% methanol (v/v) in Nanopure water and passed through reverse-phase (C-18 octadecyl) resin (Empore, 3M). The 47-mm C-18 disks were preconditioned with two 10-mL volumes of 100% ethyl alcohol (Aaper) and 10 mL of Nanopure (Nanopure Infinity) water as prescribed by the manufacturer. Retained organics were sequentially eluted off C-18 disks using 10 mL of 0.2 (volume fraction CH_3OH) methanol/water solution followed by 10 mL of a 0.5 methanol/water solution and then 10 mL of 0.8 methanol/water solution. Thus, three fractions (20, 50, and 80%) were collected from each sample and run separately on the YES or KBluc bioassays as described below. Consequently, the bioassay results may be presented in one of two forms depending on the issue being discussed: as the total estrogenic signal calculated by the sum of the responses for each fraction, or as the individual fraction's estrogenic signal.

For liquid-phase 0.7 μm filtrates, whole samples (undiluted) were applied and sequentially eluted (as described above) from C18 disks. Eluates were dried under nitrogen gas and redissolved in autoclaved Nanopure water to yield final concentration factors of 200-500x for estrogenic activity analysis. Solid-phase eluates were similarly dried under N_2 gas, resuspended with 1-2 mL of autoclaved water, then 0.7 μm glass fiber-filtered prior to analysis by bioassay.

2.4.2 Yeast Estrogen Screen (YES) Bioassay

Total estrogenic activity was measured using the YES bioassay of Routledge and Sumpter (1996) as amended by De Boever et al. (2001). The *Saccharomyces cerevisiae* strain was provided by John Sumpter of Brunel University, Oxbridge, U.K. The YES is a yeast-based *in vitro* bioassay utilizing a human estrogen receptor recombinant engineered with a beta-galactosidase reporter gene downstream of the estrogen response element. Resultant total estrogenic activity is expressed as an equivalent concentration of a known estrogenic compound – here 17α-ethinyl estradiol (EE2), an oral contraceptive.

Each sample concentrate was serially diluted across 10 wells of a 96-well micro-titer plate (Costar). Each dilution series was initiated by placing 100 μL of sample concentrate in the first well of a single row. Fifty μL was transferred to the second column and mixed with 50 μL of Nanopure water (2-fold dilution per step). The process was repeated across each row to produce a maximum dilution factor of 2^9. Fifty μL of Nanopure water that was pretreated via passage through the C-18 resin was added to wells 11 and 12 of each row to serve as (negative) process controls. The eight rows of each 96-well plate provided replicate data (n = 8) for estimation of experimental error. A standard series was developed in a similar manner with each set of measurements using concentrations of EE2 from 1.0 x 10^{-7} to 5.0 x 10^{-12} mol.

Yeast cells were grown in the Routledge/Sumpter medium to (A_{630}) 1.0 cm^{-1}. The culture was then diluted in the same medium to an absorbance (A_{630}) of 0.133 cm^{-1}, and 150 µL of the diluted suspension was added to each well of the 96-well plate (total volume 200 µL). The resultant A_{630} value in each well was then about 0.10 cm^{-1}. Plates were incubated for 24 hours at 32°C for growth of *S. cerevisiae* and estrogen-dependent expression of *lacZ*. At that point, 50 µL of cycloheximide/CPRG (chlorophenol red β-D-galactopyranoside) solution consisting of 3 mL of autoclaved Nanopure water, 2 mL of 10 mg/mL cycloheximide, and 200 µL of 10 mg/mL CPRG was added to each test well. Following an additional 24-hour incubation at 32° C for β-galactosidase-dependent color development, absorbance was measured at 570 nm (β-galactosidase activity) and 630 nm (turbidity). The contribution of cell-dependent light scattering to A_{570} measurements was determined by measuring the ratio of A_{570}/A_{630} (here defined as R) in the negative control wells. β-galactosidase activity was then corrected to $A_{570} - R \times A_{630}$. Dose-response curves were plotted for environmental samples and the positive (EE2) control.

2.4.3 T47D-KBluc (KBluc) Bioassay

The KBluc cell line bioassay developed by Wilson et al. (2004) was used on a subset of samples due to its high operational costs as a second *in vitro* technique to measure estrogenic activity. Cells were maintained in RPMI-1640 Medium with 10% fetal bovine serum (FBS) (Hyclone, Ogden, UT); no antibiotics were added to the media. The bioassay was conducted in 24-well plates and wells were rinsed with estrogen-free media containing 3% charcoal dextran treated FBS (Atlanta Biologicals, Lawrenceville, GA). Samples were serially diluted in triplicate across plates in estrogen-free growth media, and 50,000 T47D cells were seeded per well. Plates were incubated in 5% carbon dioxide (CO_2) for 48 hours at 35°C. Subsequently, cells were harvested using lysis buffer of which 100µL from each well of the lysed cell solution was collected and transferred to a 96-well luminometer plate. Luciferase activity was quantified using an Analyst AD Plate Reader (Molecular Devices, Sunnyvale, CA). The positive estrogen control consisted of decline dilutions of EE2 (Sigma-Aldrich, St. Louis, MO) from 10 nmol to 1 fmol. Data was plotted as relative light units (RLU) versus EE2 concentration. A negative control plate consisting of media and cells was run concurrently with each set of environmental samples.

2.4.4 Quality Control

Sample log sheets, with unique identifiers for each sample, accompanied all samples during bioassay analysis. For data acquisition and measurements, the quality assurance/quality control (QA/QC) plan adheres to the principles listed below.

1. To assess the potential for sample extract contamination, at least one field blank and one laboratory blank (process control) were included in analyses of sample extracts from each field site.
2. To assess the potential for loss of estrogenic compounds during sample handling/processing, spike recovery samples were used during the extraction comparison experiments.
3. To assess accuracy, duplicate sample extracts were run for 5% of the samples processed.

The QA/QC program was used to assess bioassay method performance. Method performance limits for recoveries, blanks and duplicates were established during the first phase of the study. Results from samples were used to validate data. Any data that fell outside of the performance criteria were noted with permanent delineators. If more than 10% of the data from any method were invalidated for any batch of samples, the analysts and principal investigators held a meeting to identify approaches for improving method performance.

2.4.5 Updated Data Reduction Method: First Response

2.4.5.1 EC50 Method

The traditional technique to quantify estrogenic activity in environmental samples relies upon identifying the midpoint (50%, EC50) level of response in both the environmental sample and positive control (either E2 or EE2) dose response curves. In this approach, the estrogenic response of an environmental sample is converted to an equivalent concentration of the known estrogen (EE2), used in this project using:

$$EEQ = EC50EE2/(EC50_{sample} *CF)$$

where $EC50_{sample}$ is the volume fraction of the sample producing a 50% maximal response, EC50 EE2 is the concentration of EE2 that produces a 50% maximal response in the positive control dose response curve, and CF is the concentration factor of the sample extract (typically 200-500X for liquid-phase samples).

2.4.5.2 Difficulties with EC50 Method

In this project, estrogenic activity in several samples could not be determined using the traditional EC50 method due to sample toxicity inhibiting the estrogenic response. A new data reduction method was devised and was deemed the “First Response” method. The method relies upon identifying the lowest concentration of sample in the assay plate dilution series that exhibits an estrogenic response significantly above background. A statistical approach utilizing Student’s t-tests was used to determine when significant departure from baseline occurred. Using this information, the method then follows in a similar fashion to the equation above. The new method is described in “Introduction of a new method, the First Response, to measure hormonal bioassays,” by Sondra S. Teske, Patricia Orosz-Coghlan, Wendell P. Ela, and David M. Quanrud (manuscript in preparation to be submitted to peer-reviewed journal).

A description of the problem encountered with the EC50 method is described herein and is divided into two components (A-B).

A. If the maximum estrogenic response in an environmental sample is less than the EC50 of the positive control (Figures 2-7 and 2-8), then two courses were possible. In the case where a sample provides less than an EC50 level of response, some researchers chose not to quantify the response (e.g. Andersen, 1999). These sub-EC50 responses are essentially registered as equivalent to a non-detect and useful data may be unnecessarily lost. Alternatively, other researchers have calculated estrogenic activities based on using a 20% level of response (EC20), relative to the estrogen standard, or even a 10% level of response (EC10) (Legler, 2002). However, an EC20, EC10, or any EC-based calculation will suffer the same issue of possible leftward translation of the estrogen standard detailed above that can affect results. The YES bioassay may be particularly prone to this shortcoming as it has been argued that it is less sensitive to estrogens and xeno-estrogens compared to mammalian assays (Legler, 2002).

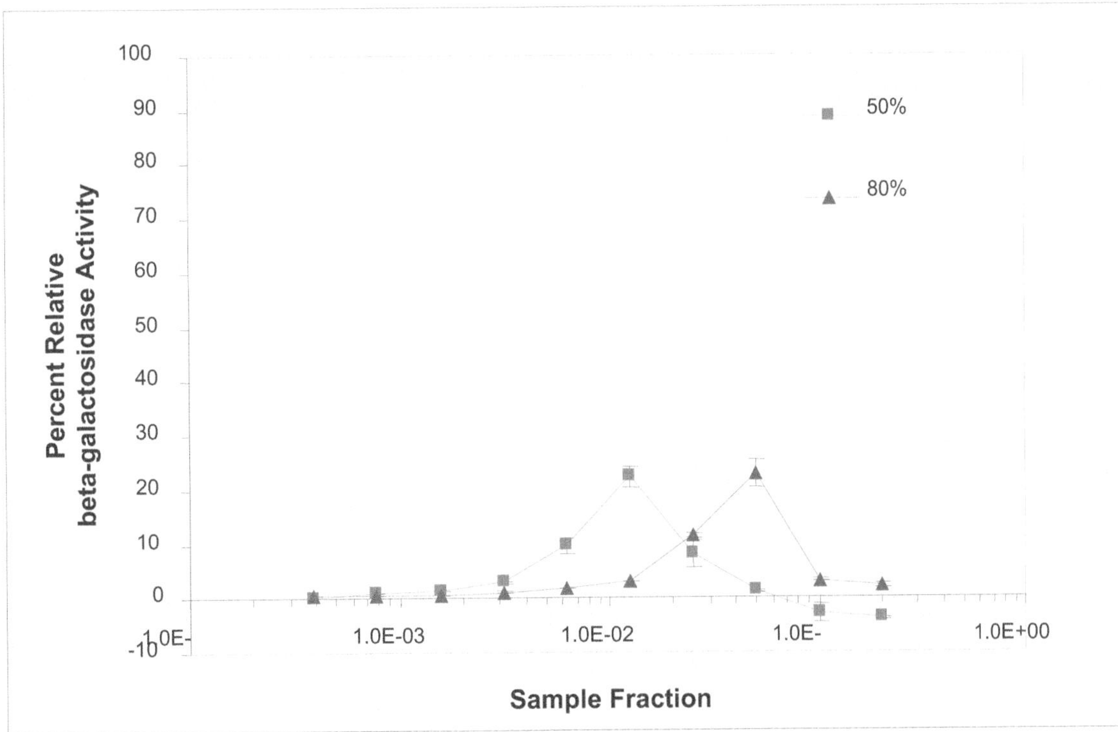

Figure 2-7. Percent Relative β-galactosidase Activity (Abs570) for the 50% and 80% Eluate Fractions in the YES Bioassay.

In Figure 2-7, environmental samples show toxic effects seen as a depression of β-galactosidase expression (Abs570) in the YES bioassay. Figure 2-8 (following) shows that toxicity suppresses the reporter gene response so that the samples only attain an EC20 (estrogenic concentration equivalent to the 20% response of the estrogen standard curve).

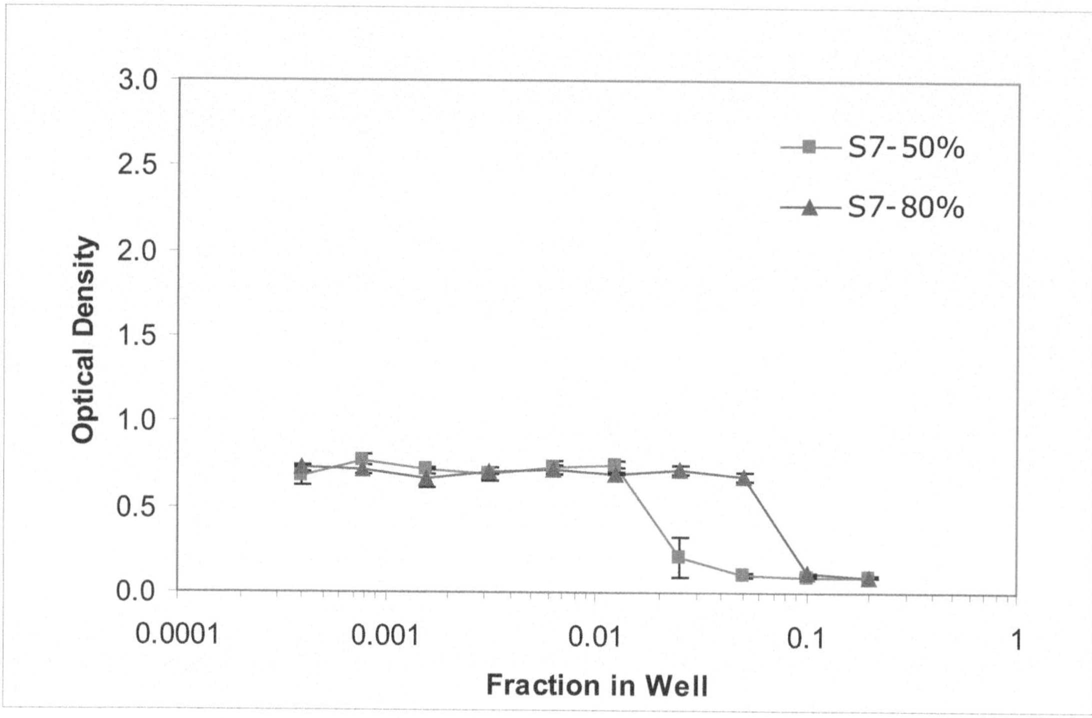

Figure 2-8. Optical Density (Abs630) Measurements Indicating Decrease in Yeast Cell Density due to Sample Toxicity.

In Figure 2-8, the YES bioassay results show a corresponding decrease in yeast cell density as measured by optical density (absorbance at 630 nm) for the first four dilutions of the 50% eluate sample, and the first two dilutions of the 80% eluate sample. Diminishment of yeast cell population depresses estrogen-linked response in Figure 2-7.

Toxic compounds in environmental samples can depress cell growth (measured by the optical density at Abs630) in a manner that correlates with increasing concentration, and accordingly suppresses β-galactosidase expression (Abs570) in the YES bioassay (Figures 2-7 and 2-8). It has been proposed that high pressure-temperature extraction methods commonly used for soils, sediment, and biosolids can promote release of toxic compounds via destruction of large organic macromolecules (humic substances) (Aerni, 2004). Some sulfur compounds also produce toxicity (Chen, 2002) in the YES bioassay. The First Response (FR) method, used by UA for this study (Section 2.4.5.3), provides a reproducible, non-subjective means to maximize data recovery in the face of cytotoxicity that usually affects bioassays near the highest concentrations of the environmental samples. Because the FR method focuses on the lower concentrations of test samples, it will avoid the impact of toxic effects that are manifested at higher concentrations (such as might be related to a non-toxic containing sample's EC50).

B. Supra maximal estrogenic responses can occur for environmental samples. That is, a sample may provide a response above the maximum response of the estrogen standard curve. These have sometimes been reported as relative induction efficiencies over 100% (Dhooge, 2006). An example of this phenomenon is shown in Figures 2-9 and 2-10 for the KBluc bioassay. The EE2 positive control attains a peak luminescence of approximately 65,000 RLU (Figure 2-11), whereas the three eluate fraction extracts of an environmental sample register from 80,000 to nearly 140,000 RLU (Figure 2-10) – significantly higher than the estrogen standard. The interpretation of an EC50 is that it defines the dose at which 50% of the population expresses the response of interest. Supra maximal responses of samples versus the standard would logically then suggest that potentially greater than 100% of the population can express the response of interest. The FR approach avoids this interpretive conundrum as it simply quantifies the point in which a response is first detected without reference to the maximum response that might be elicited.

Results cannot be compared between studies in which different EC levels are used to quantify bioassay responses (although it is frequently done) unless the logistic curves for the responses have the same slopes throughout. This is often not the case, as response curves of considerably different shapes are commonly observed in similar work and have been reported in the literature. For example, the KBluc bioassay often exhibits response curves that do not conform to a smooth logistical (or sigmoidal) curve as would be expected. Frequently, curves with differing sharpness of response and plateaus, or temporary suppressions are observed in the bioassay response to environmental samples and standards (Figures 2-9 and 2-10). These could be due to competitive and disparate effects of agonists and antagonists (Conroy, 2005; Dhooge, 2006; Silva, 2006) in the same sample affecting different steps along the complex steps of transcription-activation ligand-binding assays. Comparative results between calculations based on EC20 and EC50 of the same curve can show large disparities (Table 2-10).

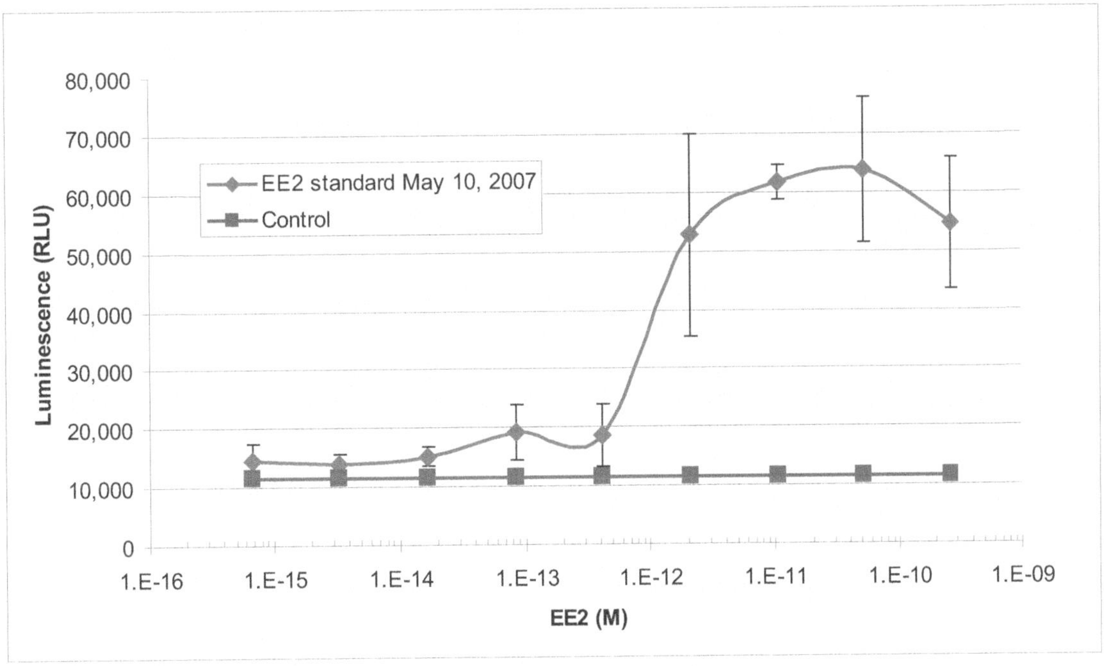

Figure 2-9. KBluc Bioassay's Ethinylestradiol Standard Curve.

As shown in Figure 2-9, the KBluc bioassay's ethynyl estradiol standard curve does not conform to logistical format due to plateau seen between 10-13 and 10-12 M concentrations. In addition, the highest dilution (10-15) is above the average background control levels plus one standard deviation.

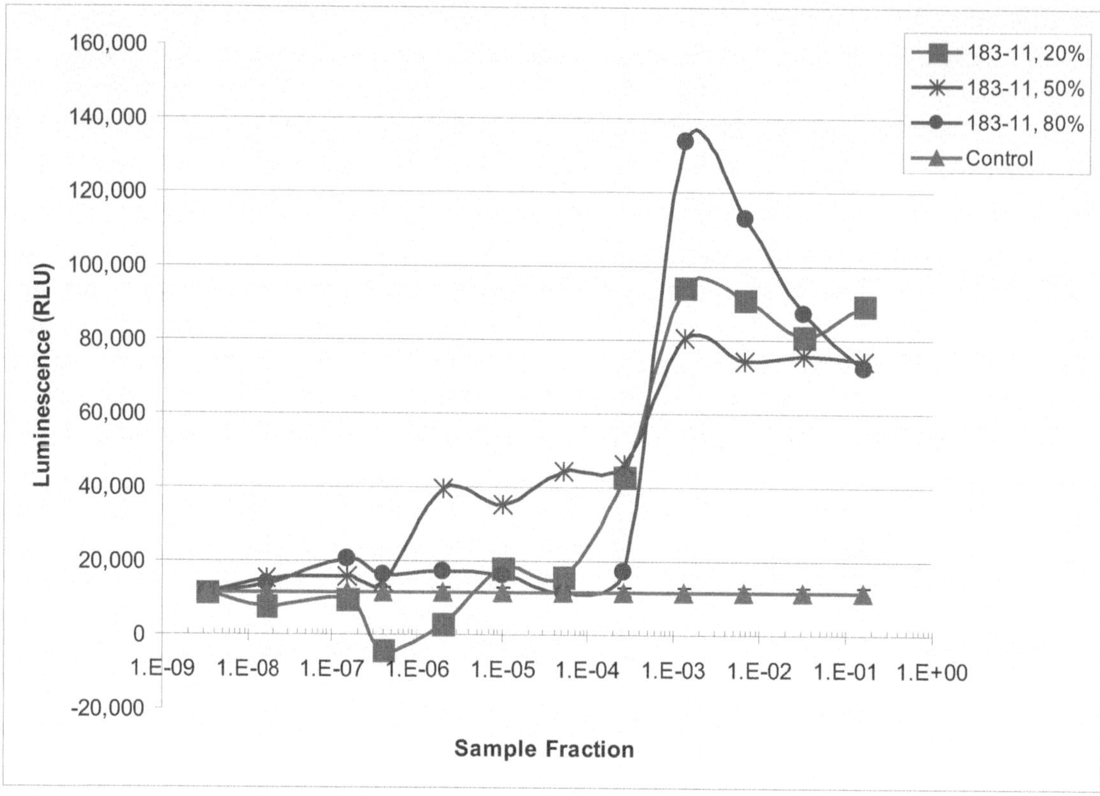

Figure 2-10. KBluc Bioassay Sample Dilution Curves.

As shown in Figure 2-10, the KBluc bioassay's sample dilution curves do not conform to a logistical format. Sample 183-11, 50% fraction exhibits two response plateaus even though dosage concentration decreases over 2 orders of magnitude: the first plateau occurs from the first dilution to the 10-3 dilution, when it drops and then holds steady from the 10-4 to 10-6 dilutions, after which it drops to control levels. Anomalous sub-control levels are also observed in the 20% fraction from the 10-5 to 10-7 levels, after which it rebounds to control levels with increasing dilutions.

2.4.5.3 First Response (FR) Method

There are several shortcomings in the use of a traditional Effective Concentration (EC50) or (EC20) protocol for analysis of environmental samples using the YES and KBluc bioassays. To overcome issues experienced during this project, a new data reduction method, deemed FR, was developed and used during the project.

The First Response method is based on identifying the most dilute sample concentration along the dose-response curve that exhibits an estrogenic response statistically above the negative control (background) response. A one-sided Student's t-test is used to determine the initial positive response of a sample or standard that is significantly higher than background. The one-sided Student t-test is appropriate for tests comparing two populations with independent means when the number of samples and controls differs and the dose-response curves vary in both slope and heteroscedasticity (noise in the response relating to dosage level). A description of the Student's t-test can be found in standard statistical texts, e.g. Bruning (1997). The degrees of freedom in the t-test calculation were determined by the number of replicates in each test group. The "First Response" value on the dose response curve is identified as the first (highest) dilution (or lowest concentration) in which the means of the test group and the negative control group are significantly different ($t > t_{critical}$) based on t-tests. In order to perform the t-test, it is necessary to select an alpha value specifying the level of statistical significance desired. The t-critical value is based on the degrees of freedom in the experiment: ($DF = [N_1 + N_2]-2$), where $N_1 + N_2$ are the number of replicates of the test group and control group, respectively, and the user-selected alpha significance value for a one-tailed test (http://www.jeremymiles.co.uk/misc/Tables/t-test.html).

An empirical procedure called Significance Level Determination (SLD) was developed as a basis to select the statistical significance values (alpha) and corresponding critical t-values appropriate for analyzing a given set of data in the YES and KBluc bioassays. The procedure is based on consideration of the magnitude of average standard error (ASE, standard deviation of each group divided by its mean) between each test group and the corresponding negative control group. The resultant ASE cutoff ranges and corresponding t values for the YES and KBluc bioassays (Tables 2-8 and 2-9, respectively) were developed through visual inspection of an existing results database, including YES and KBluc data from 130 samples through four WWTPs and 25 samples through one WWTP, respectively.

Potential presence of sample toxicity necessitated development of a parallel FR t-test analysis approach comparing cell density, measured as optical density, in tested samples versus negative controls. The "first toxicity" (FT) response was applied only for the YES bioassay; the plate reader used for the KBluc bioassay did not permit optical density readings. An (FT) response concentration was determined for each dose response sample curve obtained from the YES bioassay by measurement of optical density (light scattering) at a light absorbance of 630 nm (Abs630). In the FT analysis, alpha was set at 0.005 for all samples. The most dilute sample concentration in the plate considered to exhibit toxicity was that which showed a significant difference ($t > t$ critical) between the means of Abs630 in the test (sample) group and the negative control group. The FR analysis was accepted only when the FR-selected dilution value occurred at a lower sample concentration than the FT dilution value. An additional criteria was adopted to avoid a small incidence of false positives that was noticed initially in data analysis: the FR was accepted only when the immediately higher (less dilute) concentration also was significantly above baseline.

Table 2-8. YES Bioassay Significance Level for Samples Using the First Response Method (Degrees of Freedom (DF) = n1 (Test Group) + n2 (Control Group) – 2).

Average Standard Error (ASE)	Critical t-value for EE2 (DF = 14)	Critical t-value for sample (DF = 6)	Significance Level (α)
<0.060	5.75	10.25	0.000025
0.060<ASE<0.080	5.36	9.08	0.00005
0.080≤ ASE <0.118	4.50	6.79	0.00025
0.118≤ ASE <0.126	4.14	5.96	0.0005
0.126 ≤ ASE <0.155	3.33	4.32	0.0025
0.155 ≤ ASE <0.330	2.98	3.71	0.005
0.330 ≤ ASE	1.35	1.44	0.01

Table 2-9. KBluc Bioassay Significance Level for First Response Method (Degrees of Freedom (DF) = n1 (Test Group) + n2 (Control Group) – 2)

Average Standard Error (ASE)	Critical t value (α) (where DF = 7)	Significance Level (α)
<0.1180	4.03	0.0025
0.118≤ ASE <0.120	3.50	0.005
0.120≤ ASE <0.155	2.36	0.025
0.155 ≤ ASE <0.330	1.89	0.05
0.330 ≤ ASE	1.41	0.1

In this project, the magnitude of estrogenic response deviation from the mean was substantially higher in the KBluc bioassay than in the YES bioassay. This difference is reflected in the ranges of average standard error (ASE) and corresponding Significance Level of Determination (SLD) grouping bins for the two assays. The lower ranges of SLD for the KBluc bioassay reflect the higher "noise" in KBluc data and the lower confidence in its predictions, as the SLD directly relates to a sliding scale of alpha levels.

In order to address some of these problems in a statistically supportable method, the FR method is focused on using a rigorous statistical test that only requires data in the more highly diluted areas of the sigmoidal response curve. This avoids the arbitrariness in shifting between quantification using EC20s or EC50s with varying maximum response of the standard curve or supra-standard sample responses. In addition, the FR allows quantification of estrogenic responses for those sample curves where an EC50 approach is not possible due to sample toxicity.

Determining the background level of a zero response is critical to the determination, and some adaptations were incorporated in the FR method to transform data that did not conform to expectation that the highest dilution of the estrogen standard or sample should be essentially diluted to the zero dose response. Specifically, the expectation was incorporated that the average response of the most diluted sample (or standard) should become horizontal within one standard deviation of the negative control response. An example of where the standard EE2 curve did not meet this expectation is seen in Figure 2-9. To correct this problem when it was observed, the amount of response above background was deducted from each dilution data point to essentially shift the entire curve downward so that the highest dilution of the sample corresponded to the background response.

To illustrate the improvement in obtaining valid concentrations for low dose-response environmental samples using the FR method, estrogenic dose response curves from a set of wastewater samples were analyzed using the traditional EC20 and EC50 data reduction approach and the FR approach. Samples were obtained from a wastewater treatment plant (not part of this study) that operates two parallel secondary treatment trains (high-purity oxygen activated sludge and extended nutrient removal) followed by anaerobic digestion. Out of a total of 72 sample analyses, the percentage of samples showing a non-detect response because they were below the minimum level of positive response were 26% for the FR method, 47% for the EC20 method, and 63% for the EC50 method. The distribution of the calculated EE2-equivalent concentrations using the FR, EC20, and EC50 data reduction methods is shown in Figure 2-11. The moderate correlative relationships show equal scattering above and below the log-log association.

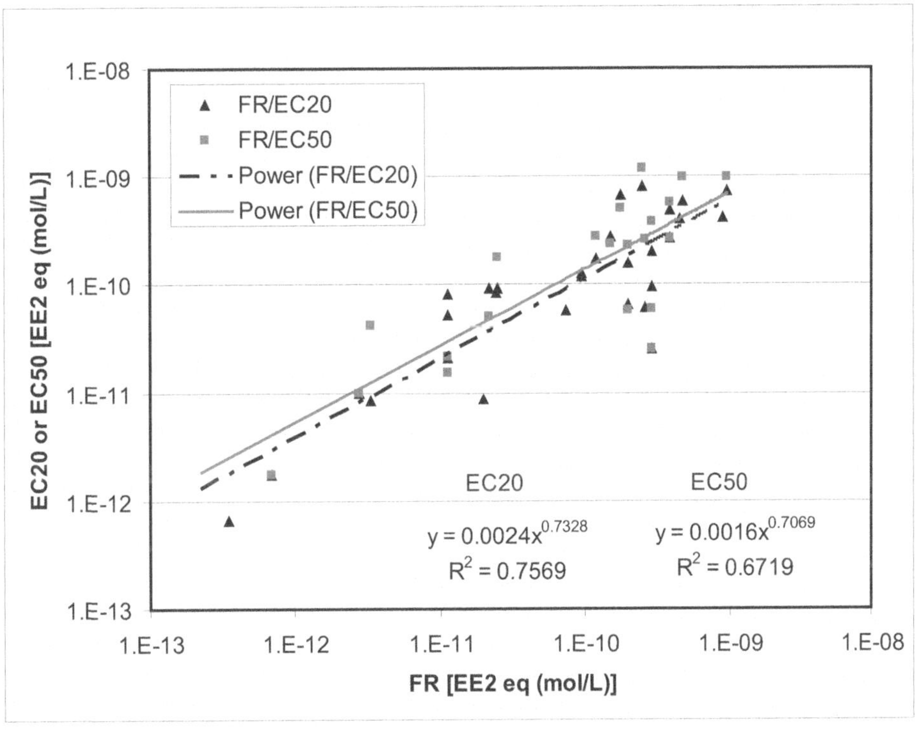

Figure 2-11. Distribution of Estrogenic Responses Obtained from the YES Bioassay and Processed Using the First Response, EC20 and EC50 Data Reduction Methods.

The revised FR data reduction method was developed as a consequence of observed inadequacies and a need to minimize the subjective components in the conventional EC method for quantitative analysis of the project bioassays. The data reduction steps in the proposed FR method follow, although in practice the modified data reduction method is implemented using a spreadsheet calculation.

1. Determine the average background absorbance as the mean value of absorbance for the negative control wells on the 96-well plate of interest (either positive control plate or environmental sample plate).
2. Determine the average relative standard deviation in absorbance of the data points defining the environmental sample response curve (n = 4) and the corresponding positive control response curve (n = 8).
3. For each response curve, calculate the absorbance level corresponding to 10 times the average relative standard deviation (step 2) above the average background level (step 1). This absorbance is the FR absorbance.
4. Determine the concentration of the first sample in which the sample absorbance minus the FR absorbance (step 3) is equal to or greater than zero. This is the FR concentration.
5. Determine the average background absorbance of the corresponding optical density (630 nm) data as the mean value of absorbance for the negative control wells on the 96-well plate.
6. Determine the average relative standard deviation in absorbance of the data points for the environmental sample's optical density data (n = 3).
7. For each optical density curve, calculate the absorbance level corresponding to 3 times the average relative standard deviation (step 6) below the average background absorbance (step 5). This absorbance is the first toxicity, FT, absorbance.
8. Determine the concentration of the first sample on the optical density curve in which the sample absorbance minus the FT absorbance (step 7) is equal to or greater than zero.
9. Select the concentration on the optical density curve which immediately precedes (is more dilute than) than the concentration determined in step 7. This is FT-1 concentration.
10. If the FT-1 concentration is less than or equal to the FR concentration (step 4), then the sample is classified as toxic. If the FT-1 concentration is greater than the FR concentration, then the FR concentration is accepted as an acceptable, valid assay result.
11. As for the IC method, the valid sample response is converted to the equivalent concentration of EE2 using:

    EE2 (equivalent) of sample = FREE2/(FR_{sample}*CF)

    where CF is the sample concentration factor.

2.4.5.4 Shortcomings of the First Response Method

The FR method overcomes a number of EC50 method shortcomings, but it also suffers from certain drawbacks attendant with use of the EC50 approach. First, the FR approach, like the EC50, still bases the entire quantification on a single point of the dose response curve rather than the entire curve. Second, it may be more prone to false positives because of its reliance on a lower response than required for the EC50 (although on the flip side of the issue, this lower sensitivity provides access to responses otherwise masked by toxicity). Specifically, concentration-dependent variability increases at the minimum and maximum responses. In a 4-parameter logistic model, the heteroscedasticity (or variance of the response due to concentration) is dampened by differential weighting to dampen responses exhibiting the highest variability. The lower signal to noise ratio related to non-specific binding responses near the lower asymptotic boundary are problematic for the First Response Method. The First Response is defined to be right outside this area of high variance. Higher t-critical values help to compensate for the uncertainty of the dose response and the Significance Level Determination method somewhat compensates for samples with high variance at low concentrations (high dilutions).

Insufficient dilution of a sample will insure that the asymptotic lower boundary of non-specific binding will not be measured, so that the most dilute sample will not qualify as the First Response. If multiple baseline contacts occur (that is, the response bounces up and down, erratically returning to baseline), a required positively-sloped response for two consecutive data points after lift-off was mandated to qualify as a First Response. In addition, use of the FR suggests that the dilution series selected for a sample be biased toward higher dilutions than might be the case if an EC50 method is used. The FR method emphasizes correctly identifying where the response curve meets the background line, whereas the EC method emphasizes utilizing a dilution series that captures the full width of the response curve from 0 to 100% response.

In applying the FR method as previously described to the KBluc bioassay results, it was observed that the method's ability to select an appropriate value for the point of first departure of the sample response from the background response was inconsistent due to considerable differences in the magnitude of variability in different KBluc results run on different days. This is not a new observation. However because the FR method considers the inherent variability in the data in determining the point where the sample response is statistically different from the background, whereas the EC50 method does not incorporate recognition of the statistical variability of the data, this batch to batch variability impacted the FR method analysis while not affecting the more subjective EC50 method.

2.5 Analytical Difficulties with Centrate Streams

Centrate samples from participating plants were consistently difficult to extract and analyze for both the UA and USGS laboratories. Based on experience, it was postulated that a colloidal phase that was not removed by centrifugation (plant or laboratory) and/or filtration was present in these samples. Activated sludge WWTPs typically use modified polyacrylamide polymer addition to thicken and flocculate sludge, and it was hypothesized that this polymer may be acting as or enhancing the postulated colloidal phase.

Samples of centrate streams were split into liquid and solid samples for analysis. In a few cases, there was not sufficient centrifugable solid material in a centrate sample to conduct a separate solids analysis and only liquid samples were analyzed. Due the high levels of colloidal material present even after filtration, solid-phase extraction media became clogged and it was generally not possible to extract a full sample. As a result, 50-100 mL aliquots of centrifuged centrate samples were diluted into 500 mL (hormones) or 1 L (pharmaceuticals/AWIs) and then processed as normal samples. Because a small amount of sample was pre-concentrated prior to analysis, MDLs for centrate samples are 5-10 times higher than for other aqueous samples.

2.6 Extract Cross Comparison Experiment

Extractions on a common set of samples were performed at USGS and UA in late April 2007. This experimentation was out of scope of the original contract but was critical to the joint interpretation of the bioassay and chemical data.

2.6.1 Introduction

In 2005, as part of an additional collaboration between the USGS and the UA, two identical sets of wastewater sludges were extracted at the UA using microwave assisted extraction (MAE) and, concurrently, at the USGS using ASE for the purpose of comparing extraction recovery efficiencies for analytes targeted in this WERF project. Results showed that ASE provided substantially higher recoveries for several target analytes in side-by-side extractions performed on a common sludge sample. Based on the discrepancies observed in that initial comparison, a more detailed comparison of MAE and ASE analyte recovery from wastewater sludges was performed. The purpose of this experiment was to compare the recovery efficiencies of MAE and ASE extraction methods for different types of sludge/wastewater samples and to determine whether or not the two extraction methods provide similar recovery efficiencies for analytes of interest.

Based on these results, the team aimed to develop a consensus approach for the analysis and interpretation of chemical and bioassay results from these methods. This was a critical step to calibrating bioassay estrogenic response to chemical constituent composition, and was particularly important because of the wide concentration and composition differences between steroidal estrogens, where individual constituents may be at μg/kg concentrations, while other non-steroidal estrogenic compounds, such as alkylphenol ethoxylates and bisphenol A are at orders of magnitude higher concentrations than the steroidal estrogens, but have orders of magnitude lower specific estrogenic activity.

2.6.2 Experimental Approach

A series of extracts were produced using MAE (UA extraction method) and ASE (USGS extraction method). Three different sludge/biosolid samples from Plant C were included in the cross comparison: centrifuged/dewatered sludge, lime stabilized sludge, and a spiked lime stabilized sludge. A duplicate lime stabilized sludge sample was included, resulting in a total of four environmental samples. A fifth test case, muffled sand, provided by USGS, was included as a blank. Each of the five sample types was prepared by USGS.

The composite samples were prepared and aliquoted at USGS. Spiking solutions and a spiking kit were shipped to UA along with instructions to minimize variations. The timing of all sample extractions was coordinated to ensure that the time between sample aliquoting and the initiation of extraction for all methods corresponded. Replicates and a matrix spike of the lime-stabilized sludge for each analysis were included for QC. Table 2-10 provides the specific samples analyzed. After extractions were performed, both labs shipped (on ice, overnight) aliquots of the resulting extracts to the other laboratory, resulting in a total of 5 MAE extracts and 5 ASE extracts for analysis in each laboratory. At UA each extract was loaded onto a C18 disk and eluted using 20%, 50%, and 80% MeOH, resulting in three discrete sample fractions that were individually evaporated to near dryness and re-suspended in ultrapure water for estrogenic activity measurement using the YES and KBluc bioassays. Estrogenic responses (EE2 equivalents) from the three fractions were summed to provide a total estrogenic activity measurement for each sample and from each assay. The USGS did a solvent exchange step on the MAE extracts as appropriate prior to analysis for the hormones, pharmaceuticals and wastewater compounds included in the WERF project.

Table 2-10. Listing of Samples Included in the Extraction Cross Comparison Experiment.

Extraction Method	Sample Name
MAE	Centrifuged/Dewatered
MAE	Lime Stabilized
MAE	Lime Stabilized, Duplicate
MAE	Lime Stabilized, Spiked
MAE	Muffled Sand
ASE	Centrifuged/Dewatered
ASE	Lime Stabilized
ASE	Lime Stabilized, Duplicate
ASE	Lime Stabilized, Spiked
ASE	Muffled Sand

Hormone data were analyzed with a smaller set of isotopically labeled surrogates in this sub-study because certain surrogates decayed during MAE due to deuterium exchange reactions in the heated methanol solvent system. This did not affect the ASE samples; however, data from the ASE samples were treated in the same way for the sake of comparability of the two data sets. Furthermore, since deuterium exchange is a phenomenon that exclusively affects the surrogates, it does not detract from the applicability of MAE to bioassay samples that do not contain surrogates. The only estrogenic steroid affected by this issue was estrone, although a number of the androgens and progestins were treated separately.

2.6.3 Observations

Chemical data for estrogenic compounds and from the wastewater indicator and hormone analyses are compiled in Table 2-11. Recoveries of estrogenic steroids compare well between the two extraction techniques. EE2, Estriol (E3), equilin, equilenin, and mestranol were not observed in unspiked sludges by either method. Low levels of diethylstilbestrol and 17α-estradiol were observed in MAE samples but not ASE samples, however they were below nominal detection limits so this did not create any inconsistency. Estrone (E1) and Estradiol (E2) both were observed in all three unspiked samples. Within methods, variability for these compounds was 12.2% or less in replicate samples, and variability between ASE and MAE techniques was somewhat higher (2-49.6%), but at levels within a factor of five of detection limits this represents good reproducibility. Recoveries of estrogens in the spiked ASE samples were acceptable with estriol (200%) being somewhat elevated. The MAE spiked sample appeared to have been contaminated with high levels of E1 and E2; equilin and equilenin had low recovery, but the other estrogens performed well. It is likely that equilin, equilenin, and some of the androgens were not extracted well by MAE because the protocol was not initially intended or optimized for such a broad suite of analytes.

In general, the non-steroidal estrogenic compounds were present in the biosolids samples at high enough levels that concentrations spiked in were overwhelmed by ambient levels, so recovery data are not reported. Nevertheless, agreement of replicates within extraction methods was very good (4.9-28%). ASE and MAE provided comparable results for nonylphenol (NP), octylphenol (OP), and their monoethoxylates (NP1EO, OP1EO). However, ASE extracts contained substantially more nonylphenol diethoxylate (NP2EO) and beta-sitosterol. These two compounds are slightly more abundant than the lower ethoxomers. Since NP2EO (Routledge and Sumpter, 1996) and beta-sitosterol (van den Heuvel et al., 2006) are significantly less estrogenic than NP and present at similar levels, the total estrogenicity of ASE and MAE extracts is comparable (Table 2-13). Conversely, bisphenol A was observed in MAE extracts, but is known to be a poor performing analyte in the USGS 5433 method and was not observed in the ASE extracts; levels were low enough that presence or absence in an extract is not likely to significantly affect the total estrogenic activity. Furthermore, although chemical analysis methods are fairly extensive, there likely are compounds with estrogenic activity that are not measured in this study, but would be expected to occur in samples (e.g., alkylphenol ethoxycarboxylates, phthalates, phytoestrogens, certain pesticides). Therefore, the predicted estrogenic potency predicted from chemical data should be looked at as a lower bound to what may be present in a complex sample.

From the chemical data generated in this cross-comparison experiment it was concluded that although the MAE technique was less effective than ASE at recovering certain analytes, it was effective at extraction of the most potent estrogens, and for the compounds of most interest the extracts have similar enough chemical composition to justify direct comparison of data between the two techniques.

A summary of estrogenic activity measurements from the extraction cross comparison experiment is shown in Table 2-12. The KBluc and YES bioassays were performed on 20, 50, and 80% MeOH elution fractions after loading the sample extract on a C18 disk; for each sample, the EE2-EQ values shown in the Table are the summed EE2-EQs obtained from the three eluate fractions.

A comparison was made between the (bioassay) estrogenic activity measurements and the predicted estrogenic response from summation of calculated activities based on chemical data. Chemical measurements were converted to equivalent concentrations of EE2 using the conversion factors listed in Table 2-13. In all cases, measured (bioassay) estrogenic activities were less than the predicted (converted chemical data) activities. Measured estrogenic activity never exceeded 20 ng/g for the KBluc bioassay or 0.2 ng/g for the YES bioassay even in the spiked sample where the equivalent of over 700 ng of E2 was added per g dry sludge. It is possible that the complex organic matrix co-extracted from sludge/biosolids has significant inhibitory effect on estrogen response in the assays. This may occur due to binding of the estrogenic compound(s) of interest to co-extracted organics, preventing transport into the cell.

The muffled sand provided an estrogenic response in both assays and in both MAE and ASE extracts. Three additional MAE extraction experiments were performed by UA to track down the source of estrogenicity. MAE extractions were performed on methanol blanks (no solid sample), and on the muffled sand that had been further cleaned by acid washing and acid washing combined with muffling again at 550°C for five hours. Extracts from each test condition were analyzed on both bioassays. The methanol blanks showed no response in either assay, indicating the MAE extraction apparatus was not a source of estrogenicity. The acid washed sand extract showed no response in the YES bioassay but was detected in the KBluc bioassay (about 1.0E-13 EE2-EQ per g). The combined acid-washed and remuffled sand was again negative in the YES bioassay and showed a response in the KBluc bioassay of about 5X10E-14 EE2-EQ in the KBluc bioassay. Results of this additional work suggest compounds extracted from the sand were the source of estrogenic response. The measured estrogenic responses listed in Table 2-12 were not corrected for the sand-contributed estrogenic activity because that necessitates an assumption of additivity that may not be appropriate in this case. Further study in this area is warranted.

Table 2-11. Extraction Comparison Data for Chemical Analysis, Plant C.

Steriod	Sample and Extraction Method							
	MAE				ASE			
	Centrifuged, Dewatered Sludge	Lime-Stabilized Sludge	LSS-Dup	LSS + Spike	Centrifuged, Dewatered Sludge	Lime-Stabilized Sludge	LSS-Dup	LSS + Spike
Diethylstilbestrol	0	0.4	0	165.7	0	0	0	459.7
17-alpha-estradiol	0	1	1.1	100.2	0	0	0	389.6
Estrone	5.2	3.5	3.6	7,670	5.4	2.8	2.5	368.5
Equilin	0	0	0	5.6	0	0	0	324.1
17-beta-estradiol	1.5	2.3	2.1	7,770	3.1	2.3	2.1	370.8
Mestranol	0	0	0	159.4	0	0	0	357.1
Equilenin	0	0	0	61.9	0	0	0	261
17-alpha-EE2	0	0	0	141.6	0	0	0	329.9
Estriol	0	0	0	26.7	0	0	0	60.9
4-tert-octyphenol	6,330	5,850	4,200	6,790	8,200	6,900	6,450	7400
Para-nonylphenol (TOTAL)	78,000	97,000	68,000	83,000	110,000	110,000	115,000	130,000
4-cumylphenol	0	0	0	2,830	0	0	0	0
OPEO-1	640	2,560	1,890	3,610	0	780	865	1,600
NPEO1-total	147,000	240,000	209,000	260,000	210,000	300,000	290,000	340,000
OPEO-2	0	296	0	0	0	0	0	0
Bisphenol A	0	5,140	6,360	9,570	0	0	0	0
NPEO2-total	31,000	50,000	45,000	66,000	260,000	330,000	275,000	320,000
Beta-sitosterol	38,500	99,600	94,700	99,700	220,000	260,000	250,000	310,000

Table 2-12. Estrogenic Activity Results from the YES and KBluc Bioassays.

Extraction Method	Sample Name	Kbluc (EC20)	Kbluc (FR)	YES (EC20)	YES (FR)	Summed Activity
MAE	Centrifuged/dewatered	1.09E+00	1.15E-02	1.42E-01	8.53E-04	4.52E+00
MAE	Lime stabilized	2.40E-03	3.76E-03	4.00E-01	1.37E-01	7.00E+00
MAE	Lime stabilized, duplicate	1.63E+00	2.97E+00	1.09E-01	1.50E-01	6.31E+00
MAE	Lime stabilized, spiked	2.05E+00	1.64E+00	2.71E-02	6.79E-01	6.93E+03
MAE	Muffled sand	3.97E-02	1.89E-02	6.85E-02	4.13E-02	-
ASE	Centrifuged/dewatered	3.64E-01	2.85E-02	7.68E-02	2.07E-01	7.33E+00
ASE	Lime stabilized	1.21E-01	1.40E-01	2.61E-01	2.59E-01	6.43E+00
ASE	Lime stabilized, duplicate	2.45E-02	6.84E-04	9.79E-02	2.76E-02	5.63E+00
ASE	Lime stabilized, spiked	1.73E+00	1.72E+01	6.59E-01	2.31E-01	1.43E+03
ASE	Muffled sand	7.97E-01	1.80E+00	1.78E-02	1.02E-02	-

Note: estrogenic activity results from the YES and KBluc bioassays (EE2-EQ, g/g solid) for the extraction cross-comparison experiment were calculated using two data reduction techniques: EC20 and the revised First Response method. Summed activity represents the calculated EE2-EQs based on the Model of Concentration Addition using the individual chemical concentration measurements multipled by their estrogenic potency factors, relative to EE2 (Table 2-13).

Table 2-13. Compiled Conversion Factors for Selected Estrogenic Compounds in the YES Bioassay (All are Relative to EE2).

Compound	Potency Relative to EE2
Estrone	0.319
17-alpha-estradiol	0.84
17-beta-estradiol	0.84
Estriol	0.002
17-alpha-ethinyl-estradiol	1
4-tert-octylphenol	0.00036
para-nonylphenol	0.00001
4-cumylphenol	0.000001
OPEO-1	0.00001
NPEO1	0.000001
OPEO-2	0.00001
Bisphenol A	0.000068
NPEO2	0.000001
Beta-sitosterol	0.000001

2.7 Instantaneous Load Calculations

One of the objectives of this project was to calculate a mass balance of known estrogenic compounds and the total estrogenic activity throughout the four study plants. Two parameters necessary for quantifying a mass balance and the removal capability of various treatment processes are: an accurate measure of the concentration of the target analytes into and out of each of the critical unit processes; and an accurate measure of the flows and solids loadings for each sample point.

Data on flows and solids loadings based on measured data were provided by the study plants to calculate the mass balances of estrogenicity and individual TOrCs. The four study plants have detailed monitoring programs throughout their facilities which provide information on the operating mode and efficiency of each unit process. However, in several instances the mass balance of flow and solids (using total suspended solids (TSS)) across the unit operations and the interconnected network of flows, sidestreams and recycle streams for the plants was not easily closed. An assessment of plant data showed that in some instances, not all of the process streams sampled were metered and some erroneous flow and mass measurements were identified. Thus, the solids mass balance around certain unit operations did not always show closure. As a results flow splits were estimated as accurately as possible. The basis of flows and solids loadings is detailed in this section.

In addition to error associated with plant data, the SRT through the solids treatment trains as well as discontinuous production of certain solids streams (e.g. dewatered sludge) prohibited flow weighted, 24-hour composite sample collection, further contributing to uncertainties regarding mass balance calculations. Based on this information, it was determined that the term "mass balance" was not appropriate for the calculations. It is more accurately described as calculations of the loads of estrogenicity and estrogenic and other target compounds at the time of sample collection. This provides a snapshot or the "instantaneous load," for each sample and sample collection date.

2.7.1 Approach

To calculate the instantaneous loads of TOrCs and estrogenic activity, the analytical results were multiplied by the solids loading, total suspended solids (TSS), for each sample point (e.g. tons per day) to obtain the daily load of each compound, presented in grams per day (g/day).

For results of chemical analyses, the concentration of each target analyte, in nanograms per gram (ng/g) for solid samples and nanograms per liter (ng/L) or micrograms per liter (ug/L) for liquid samples, was multiplied by the flows and solids loadings values for the plant to calculate the instantaneous load of that analyte in g/day.

For results of bioassay analyses, the sum of each eluate fraction was multiplied by the flows and solids loadings for the plant to calculate the instantaneous loads of estrogenicity in mol/day in EE2 equivalents. These loadings were then converted to g/day based on the molecular weight of EE2 (296.4 g/mol).

2.7.2 Individual Plant Flows and Solids Loadings

2.7.2.1 Plant A Flows and Solids Loading

Flow data was provided by Plant A and solids were analyzed by UA. Several adjustments were made to these data based on atypical results for solids concentration of the thickened sludge as well as the flow split out of the dewatering centrifuge.

Due to inaccurate metering, flow values for Plant A for the dewatered sludge and the centrate recycle stream, sample points 3 and 4 respectively, were calculated based on measured TSS concentrations in a mass balance around the centrifuge assuming solids were conserved. Based on these calculations, the average flow of the dewatered sludge out of the centrifuge ranged from 15-20%, which was a conservative flow based on typical dewatered sludge flows of approximately 10%. The centrate stream flow was the balance of the digested sludge flow less the dewatered sludge flow.

For all sample periods, the solids values for thickened sludge have been calculated based on the volatile solids reduction typically seen in aerobic digesters, which range from 35-50% depending upon the digester liquid temperature and sludge age (Metcalf and Eddy, 2003). The solids values for thickened sludge at Plant A are based on an assumed 45% VSS reduction following digestion. Additionally, solids concentrations for the first sampling period were not analyzed, so the values were calculated based on the average concentrations of the latter three sampling periods. With these adjustments, the flows and solids loadings values around the centrifuge are within normal ranges.

Table 2-14 provides the data used in calculating the instantaneous loads.

Table 2-14. Plant A Flows and Solids Loadings.

Sample Location	Solids Load (tons/day)	Solids Load (g/day)	Flow (GPD)	Flow (L/day)
March 2006				
Thickened Sludge (Primary & Secondary)	4.64	4,209,101	34,560	130,810
Aerobically Digested Sludge	3.38	3,063,822	34,560	130,810
Dewatered Sludge	3.35	3,036,364	5,760	21,802
Centrate Recycle Stream from Dewatering Process	0.03	26,168	28,800	109,008
July 2006				
Thickened Sludge (Primary & Secondary)	9.36	8,491,498	72,000	272,520
Digested Sludge	6.87	6,235,039	72,000	272,520
Dewatered Sludge	6.75	6,124,777	14,400	54,504
Centrate Recycle Stream from Dewatering Process	0.03	27,001	57,600	218,016
October 2006				
Thickened Sludge (Primary & Secondary)	15.46	14,022,926	100,800	381,528
Digested Sludge	11.09	10,065,064	100,800	381,528
Dewatered Sludge	10.62	9,632,038	15,840	59,954
Centrate Recycle Stream from Dewatering Process	0.16	141,000	84,960	321,574
January 2007				
Thickened Sludge (Primary & Secondary)	7.77	7,045,689	59,040	223,466
Digested Sludge	5.70	5,171,576	59,040	223,466
Dewatered Sludge	5.75	5,216,556	8,640	32,702
Centrate Recycle Stream from Dewatering Process	0.04	33,175	50,400	190,764

2.7.2.2 Plant B Flows and Solids Loading

Flow and solids loadings data were provided by Plant B for use in this study. A closed flow balance was not achieved for the plant using the recorded data which suggests either errors in the metered flow measurements or internal recycle flows that were not metered or monitored (such as wash down water or backwash recycles being internally returned without metering etc.). Based on analysis of plant data it was determined that the following measurements are reasonably accurate: plant influent flow measurement; activated sludge influent flows and loads; centrate and thickener recycle flows and loads; and final effluent flow. Other measurements of flows or loads in the liquid process treatment train appeared to have varying degrees of inaccuracy, in which case the flows and loads were calculated.

A solids mass balance for the digestion and dewatering processes was also assessed for Plant B. This mass balance appeared to close to within 7% accuracy based upon reported historical values. One significant source of uncertainty in the reported data concerned a flow split of thickened sludge from the gravity thickener underflow to the gravity belt thickeners. A small amount of thickened primary sludge was added to the secondary waste activated sludge before it was thickened in the gravity belt thickeners but the mass was not recorded. An estimate of this mass was made using the data from the waste activated sludge flow metering and the gravity belt thickener influent.

Annual averages for the year 2005 were used to calculate the instantaneous loads at Plant B. Table 2-15 provides the data used in calculating the instantaneous loads.

Table 2-15. Plant B Flows and Solids Loadings, 2005.

Sample Location	Solids Load (tons/day)	Solids Load (g/day)	Flow (MGD)	Flow (L/day)
Primary Influent	122.4	111,067,529	155.00	586,675,000
Primary Effluent	49.9	45,227,267	151.00	571,535,000
Secondary Effluent	4.7	4,307,767	152.00	575,320,000
Primary Unthickened Sludge	100.1	90,842,726	3.90	14,761,500
Secondary Unthickened Sludge	70.9	64,334,434	2.20	8,327,000
Thickened Sludge (Combined Primary & Secondary)	83.7	75,917,910	0.40	1,514,000
Anaerobically Digested Sludge (Conventional Digesters)	30.9	28,001,590	0.30	1,135,500
Anaerobically Digested Sludge (Egg-shaped Digesters)	37.8	34,269,406	0.30	1,135,500
Acid Phase Digested Sludge	17.0	15,422,141	0.10	378,500
Methane Phase Digested Sludge	12.0	10,898,811	0.10	378,500
Dewatered Sludge	84.9	77,037,869	0.09	353,898
Centrate Recycle Stream from Dewatering Process	2.8	2,544,509	3.91	14,810,787
Tertiary Pelletized Sludge	45.0	40,823,313	0.05	189,250
Composted Sludge	29.0	26,308,357	0.03	121,120

2.7.2.3 Plant C Flows and Solids Loading

Flow and solids loadings data were provided by Plant C for use in this study. The mass balance program has a significant level of sophistication and can be modified to reflect a long-term average mass balance or provide details for a specific day. A review of the basis of mass balance data for this plant, which concluded that the calculations are reasonable and that flow and solids loadings data should be used in calculating the instantaneous loads at Plant C.

Not all process flow streams are metered at this plant, thus some of the flows and solids loadings were calculated. The sample points in this study and the means by which their flows and solids loadings were generated are:

- Primary Waste Sludge: flow (metered) and solids mass (calculated)
- Secondary Waste Sludge: flow (metered) and solids mass (calculated)
- Waste Sludge from the Nitrification/Denitrification Process: flow (metered) and solids mass (measured)
- Dewatered Sludge: solids mass (calculated)
- Lime Stabilized Sludge: solids mass (calculated)

Five-day averages were used to calculate the instantaneous loads at Plant C to avoid influences of variable solids mass loadings data. Table 2-16 provides the data used in calculating the instantaneous loads.

Secondary treatment is divided into two plants, East and West. In December 2005 the Secondary Waste Sludge sample was collected from the West Secondary Plant, prior to blending with the waste sludge from the East Secondary Plant. In order to make comparisons between the waste sludge streams and the dewatered sludge, it was assumed that analytical results would be comparable for both plants so the flows and solids loadings from both the West and East Secondary Plants were combined prior calculating the solids loadings for that sample point. This provides a better estimate of the total loadings of secondary waste sludge to treated solids at Plant C.

Table 2-16. Plant C Flows and Solids Loadings.

Sample Location	Solids Load (tons/day)	Solids Load	Flow (MGD)	Flow (L/day)
December 2005				
Primary Waste Sludge	151.1	137,075,614	-	-
Secondary Waste Sludge	130.9	118,750,482	-	-
Nit/Denit Waste Sludge	10.4	9,434,721	-	-
Dewatered Sludge	248.0	224,981,816	-	-
Centrate Recycle (L)	-	-	1.84	6,964,400
Centrate Recycle (S)	31.2	28,304,164	-	-
Lime Stabilized Sludge	272.7	247,389,279	-	-
July 2006				
Dewatered Sludge	326.2	295,923,662	-	-
Centrate Recycle (L)	-	-	1.92	7,267,200
Centrate Recycle (S)	46.5	42,184,090	-	-
Lime Stabilized Sludge	376.0	341,101,462	-	-

For this plant, all samples were treated as solids samples with the exception of centrate samples. As described in Section 2.5, centrate samples from participating plants were consistently difficult to extract and analyze. Centrate samples were separated into two aliquots for both liquid and solid analysis and are reported as such. In the cases when there was not sufficient centrifugable solid material in a centrate sample to conduct both analyses it is noted in the data tables. Also note that in many cases, the total mass or volume of centrate solid or liquid samples was insufficient to permit more than a single chemical analysis. In all cases, hormone analyses were chosen as the preferred analysis for mass- or volume-limited samples.

2.7.2.4 Plant D Flows and Solids Loading

The flows and loadings data for Plant D were taken from the Monthly Performance Reports provided by the plant (March 2006, June 2006, September 2006, and December 2006). Table 2-17 provides the data used in calculating the instantaneous loads. The flows and loading data for digested sludge, provided in the Monthly Performance Reports, are estimates.

Table 2-17. Plant D Flows and Solids Loadings.

Sample Location	Solids Load (tons/day)	Solids Load (g/day)	Flow (MGD)	Flow (L/day)
March 2006				
Primary Sludge (Unthickened)	545.0	494,415,683	2.65	10,030,250
Thickened Waste Activated Sludge (TWAS)	146.5	132,902,564	0.50	1,892,500
Digested Sludge	281.5	255,372,504	3.15	11,922,750
Centrate Recycle Stream from Dewatering Process	20.5	18,551,928	2.30	8,705,500
June 2006				
Primary Influent	520.0	471,736,065	332.00	1,256,620,000
Primary Effluent	701.5	636,390,095	343.00	1,298,255,000
Secondary Effluent	32.0	29,029,912	329.00	1,245,265,000
Primary Sludge (Unthickened)	498.0	451,778,001	2.47	9,348,950
Waste Activated Sludge (Unthickened)	231.5	210,013,267	9.10	34,443,500
Thickened Waste Activated Sludge (TWAS)	145.0	131,541,787	0.50	1,892,500
TWAS Centrate	76.5	69,399,633	8.80	33,308,000
Digested Sludge	280.5	254,465,320	2.98	11,279,300
Centrate Recycle Stream from Dewatering Process	21.3	19,277,676	2.60	9,841,000
September 2006				
Primary Sludge (Unthickened)	418.5	379,656,814	2.51	9,500,350
Thickened Waste Activated Sludge (TWAS)	145.0	131,541,787	0.48	1,816,800
TWAS Centrate	76.0	68,946,040	7.60	28,766,000
Digested Sludge	281.0	254,918,912	3.09	11,695,650
December 2006				
Primary Influent	555.5	503,941,123	331.00	1,252,835,000
Primary Effluent	627.0	568,804,832	342.00	1,294,470,000
Secondary Effluent	28.0	25,401,173	329.00	1,245,265,000
Primary Sludge (Unthickened)	431.5	391,450,215	2.23	8,440,550
Waste Activated Sludge (Unthickened)	199.0	180,529,763	7.90	29,901,500
Thickened Waste Activated Sludge (TWAS)	132.5	120,201,978	0.38	1,438,300
TWAS Centrate	58.0	52,616,715	7.50	28,387,500
Digested Sludge	281.0	254,918,912	3.09	11,695,650
Centrate Recycle Stream from Dewatering Process	19.5	17,644,743	2.30	8,705,500

CHAPTER 3.0

RESULTS AND DISCUSSION

3.1 Introduction

Results and corollary discussion from the study are organized into three major groups. First, the general discussion of the chemical and biological data reduction approaches is presented. This is followed by presentation and discussion of results for each plant, including: calculated instantaneous loads for hormones, alkylphenolic compounds, and bioassays; chemical analysis data reduction results and discussion; followed by discussion of the data reduction results of biological analysis and the Model of Concentration Addition approach. The final subsection discusses non-estrogenic TOrCs (e.g. select pharmaceuticals).

Results for all chemical analyses and biological analyses discussed in this section can be found in the USGS web publication (Furlong et al., 2010) (http://infotrek.er.usgs.gov/pubs/).

3.2 Chemical and Bioassay Data Reduction

3.2.1 Chemical Analysis

The primary goal of this project was to assess the fate and transport of estrogenicity through unit transport processes in WWTPs. As such, the interpretation of chemical data is broken up by compound class. This delineation does not directly correspond to the separation of compounds of interest into three separate chemical analyses per matrix.

The first section will examine the behavior of estrogenic compounds during treatment. Data were used from both the hormone analyses as well as the AWI analyses. These are the data that will be directly compared to the bioassay data based on YES bioassay results. Compounds include eight estrogenic steroids plus diethylstilbestrol, a stilbene from the hormone analysis, as well as alkylphenol ethoxylates (APEOs) and several other compounds from the AWI schedule (Table 2-5). The estrogenic EDCs investigated during this study likely have the most potent biological effects at environmentally relevant concentrations of any of the compounds examined. It is well documented that steroidal hormones can induce feminization in fish and other aquatic organisms at concentrations of 1 ng/L or less (Routledge et al., 1998). The APs, APEOs, and other synthetic compounds have potencies that may be 1,000 times or less than steroidal estrogens. However, they operate by the same estrogen receptor (ER)-binding mechanism as the natural and synthetic estrogens, and the effects of all ER agonists may be additive or even synergistic. Since many are ubiquitous man-made chemicals (or degradates) which occur 1,000 or more times greater concentration than the steroids, their contribution to total estrogenicity cannot be discounted. This is especially true in solids, to which they partition preferentially. The discussion of fate and transport of estrogenic compounds will focus first on their removal from the liquid phase, since discharge of secondary effluent to surface waters is their most direct route to the aquatic environment. But removal from the liquid phase does not necessarily constitute transformation and reduction in estrogenicity because these compounds could either be sorbed to particulate matter and still active in the biosolids, or transformed into metabolites that remain estrogenic. Therefore, the continuing discussion will assess a) whether compounds are transformed, or simply transferred into the solid phase, and b) the extent to which various unit processes are effective at reducing concentrations of estrogenic compounds.

The next set of compounds to be considered is non-estrogenic EDCs. The compounds discussed here are the remainder of compounds in the hormone analysis not discussed in the previous section. Although assays do exist to assess biological activity of these compounds, they were not employed in this study. Nevertheless, useful information about potential androgenicity can be gleaned from this data set. Progesterone mimics (i.e., progestins) are reported in the data tables, however, the number analyzed was small and analytical performance was more variable than for the estrogens and androgens, therefore interpretation was limited.

Third, an important contribution of this project to the base of scientific knowledge beyond the scope of the original proposal is information regarding the behavior of non-EDC pharmaceutically active compounds. These compounds have known biological activity, but in general it is less certain whether they might induce environmental effects at dosages found in WWTP streams. Still, there is considerable interest in understanding their environmental behavior as they potentially could have effects on aquatic biota. Therefore, a discussion of pharmaceutical fate and transport will follow.

It is important to note that the absence of diethylstilbestrol (DES) would be expected. DES is a synthetic estrogen with limited use due to complications when administered to pregnant women. It has other therapeutic uses, but is rarely prescribed. However, DES was the largest component of the estrogenic signal in digested solids for one of the plants (A). Due to improvements in GC/MS/MS analysis of DES over the course of the study, and the lack of a likely major source term for DES, the confidence in this conclusion and detections at the other study plants is less than for the other hormones and estrogenic AWIs. Briefly, a change in ion selection for a more specific MS/MS transition occurred after processing of all samples for this project. Although it cannot be absolutely determined after the fact that the measured DES concentrations are analytical artifacts, subsequent USGS analysis of many biosolids samples from other projects has failed to yield a single detection of DES. Therefore, we present the data generated according to the method used in 2006 and 2007 with some reservation.

3.2.2 Biological Analysis

Table 3-1 lists the top 16 estrogenic TOrCs, in terms of their contribution to total estrogenicity, that were detected in samples from this study. The first five compounds are steroidal hormones, both natural and synthetic; the next eight compounds are alkylphenols; and the last four compounds are other prominent chemicals frequently detected in sludges and biosolids. The estrogenic potency of each compound, relative to EE2, is given, based on literature values reported using the YES bioassay and the KBluc bioassay (or the E-Screen bioassay). There are few potency factor values published in the literature. For compounds without currently published factors, values were taken from one of two sources: personal communication from Dr. F. Leusch, who has collaborated with Dr. V. Wilson, who developed the KBluc bioassay, or published potency factors for the E-Screen bioassay. The E-Screen bioassay is also a human breast cancer cell based assay, although it uses MCF7, rather than the T47D cell line of the KBluc bioassay. The latter surrogate approach is admittedly imperfect;

however it was deemed the most defensible for cases where neither published nor unpublished but credible KBluc values could be obtained.

In brief, each sample was extracted with methanol using MAE; the methanol extract was passed through a C18 disk (3M) and the disk sequentially eluted using 20%, 50%, and 80% MeOH to provide three discrete eluate fractions for analysis on the YES bioassay. Estrogenic activity results are expressed as EE2 equivalents concentration (EE2-EQs, mol/g or mol/L) and were developed using the revised FR data reduction method as described in Section 2.4.5. Primary influent, primary effluent, and centrate recycle were separated into liquid and solid components by centrifugation. Supernatants were concentrated using C18 and differentially eluted as above; resultant EE2-EQs are reported as mol/L. Solids collected from centrifugation were processed using MAE and C18; resultant EE2-EQs are reported as mol/g.

The efficacy of the four solids stabilization processes was analyzed by comparing the amount of estrogenicity present in the solids before and after stabilization. Estrogenicity was determined by two different methods: 1) summation of each compound's measured concentration multiplied by its EE2-equivalent potency factor (Table 3-1) and 2) total estrogenic activity measured using the YES bioassay. In the following sections, plant performance in removing estrogenic compounds is evaluated based on comparison of instantaneous estrogenic mass fluxes across treatment processes.

Mass fluxes at each sampling point were calculated using the Model of Concentration Addition. This model was first proposed by Fraser (1872), and more fully described by Loewe (1926). It proposes that in a mixture, if individual chemicals structurally act in a similar way, each component can substitute for any other component at equi-effective concentrations with the same net result. This concept implies that most estrogens and xenoestrogens act on the estrogen receptor similarly. Research has corroborated that individual compounds that test below minimal detectable levels can in combination produce significantly measurable effects (Silva, 2002). Mathematically the model simply assumes that the estrogenic contribution of each individual compound is linearly additive, so that a summation of all compound's concentration multiplied by their respective EE2-equivalent potency factors (Table 3-23) is the expected total estrogenicity of the sample (in EE2 equivalents). The instantaneous estrogenic mass flux is then the total (summed) estrogenic concentration of the sample times the flow rate or solids loading rate at the sample point.

It is important to note that although the term "instantaneous mass flux" might be more appropriate in the context of this report given adoption of the term "instantaneous load" (Section 2.7), the term "mass flux" was used for these analyses for ease of discussion.

Table 3-1. Top 16 Estrogenic Compounds Detected in this Study and Their Potency Factors, Relative to EE2.

Compound Name	YES Potency, relative to EE_2 [molEE2/mol]	YES Reference	KBluc Potency, relative to EE_2 [molEE2/mol]	KBluc Reference
17α-ethinylestradiol	1	Aerni, 2004	1.0000000^{ES}	Leusch, 2010†
17α-estradiol	0.84	Sanseverino 2005	0.1000000^{ES}	Soto, 1995*
17β-estradiol	0.84	Aerni, 2004	2.8100000^{ES}	Leusch, 2010†
Estrone	0.319	Aerni, 2004	0.0600000^{ES}	Leusch, 2010†
Estriol	0.002	Aerni, 2004	0.0135^{ES}	Leusch, 2010†
4-n-Octylphenol	0.00036	Routledge, 1996	0.000105^{ES}	Leusch, 2010†
4-tert-Octylphenol	0.00036	Routledge, 1996	0.000054	Leusch, 2010†
4-Octylphenol monoethoxylates	0.00001	Estimated‡	0.00000065	Estimated‡
4-Octylphenol diethoxylates	0.00001	Estimated‡	0.00000081	Estimated‡
4-Nonylphenol	0.00001	Routledge, 1996	0.00002884^{ES}	Fang, 2000*
4-Nonylphenol monoethoxylates	0.000001	Env Canada 2001	0.00000034	Estimated‡
4-Nonylphenol diethoxylates	0.000001	Routledge, 1996	0.00000042	Estimated‡
Diethylstilbestrol	0.924	Folmar, 2002	1	Wilson, 2004+
Bisphenol A	0.000563	Matsumoto, 2004	0.000006^{ES}	Leusch, 2010†
Benzophenone	0.000168	Kawamura2003; Kunz,2006	0.001	Kawamura2003*
Diethylhexyl phthalate	0.000021	Petrovic, 2004	0.00000001	Okubo, 2003*

Notes: ‡ = Estimated means YES-based estrogenic potencies for ratios between NP/NP1EO and NP/NP2EO were applied to OPEO species and for KBluc NPEO and OPEO species, † = KBluc potency based on unpublished data provided by F. Leusch via personal communication, + = Potency based on published KBluc bioassay data, ^{ES} = Potency based on E-SCREEN bioassay data (see text for explanation).

3.3 Plant A

3.3.1 Instantaneous Load: Hormones, Alkylphenolic Compounds, and Bioassays

Instantaneous loads calculations for all sample dates for steroid hormones and alkylphenolic compounds results are provided in Tables 3-2 and 3-3, respectively. Table 3-4 shows the instantaneous loads results for the YES bioassay.

Table 3-2. Plant A: Instantaneous Loads Results (g/day), Hormones.

Sample Location	Matrix	diethylstilbestrol	cis-androsterone	epitestosterone	17-alpha-estradiol	stanolone	androstene-3,17-dione	estrone	17-beta-estradiol	testosterone	equilin
March 2006											
Thickened Sludge (Combined Primary and Secondary)	Solid	ND	0.13	ND	ND	0.18	0.14	0.027	0.0091	ND	ND
Aerobically Digested Sludge	Solid	ND	E 0.074	ND	ND	ND	E 0.064	0.034	ND	ND	ND
Dewatered Sludge	Solid	ND	0.11	ND	ND	ND	0.068	ND	ND	ND	ND
Centrate Recycle Stream	Liquid	< 0.00044	0.0067	< 0.0022	< 0.00044	< 0.00044	0.0025	0.0022	< 0.00044	< 0.00044	< 0.0022
Centrate Recycle Stream	Solid	ND	E 0.065	ND	0.00020	0.00085	0.0021	0.0060	0.00025	ND	ND
July 2006											
Thickened Sludge (Combined Primary and Secondary)	Solid	ND	0.025	ND	ND	ND	ND	0.037	ND	ND	ND
Aerobically Digested Sludge	Solid	0.071	0.14	ND	ND	ND	ND	ND	ND	ND	ND
Dewatered Sludge	Solid	ND	ND	ND	ND	ND	ND	0.21	ND	ND	ND
Centrate Recycle Stream	Liquid	D-R									
Centrate Recycle Stream	Solid	insufficient solids to centrifuge									
October 2006											
Thickened Sludge (Combined Primary and Secondary)	Solid	0.16	0.25	ND	ND	ND	0.096	0.047	0.035	ND	ND
Thickened Sludge (Combined Primary and Secondary) (Duplicate)	Solid	ND	0.19	ND	ND	ND	0.097	ND	ND	ND	ND
Aerobically Digested Sludge	Solid	0.48	ND	ND	ND	0.11	0.090	0.16	0.22	ND	ND
Dewatered Sludge	Solid	ND	0.067	ND	ND	ND	0.060	ND	ND	ND	ND
Centrate Recycle Stream	Liquid	< 0.0013	0.011	< 0.0064	0.0044	0.0023	0.0016	0.0089	0.013	< 0.0013	< 0.0064
Centrate Recycle Stream	Solid	insufficient solids to centrifuge									
January 2007											
Thickened Sludge (Combined Primary and Secondary)	Solid	ND	0.16	ND	ND	ND	0.10	ND	ND	ND	ND
Aerobically Digested Sludge	Solid	ND	0.12	ND	ND	ND	0.12	0.028	ND	ND	ND
Dewatered Sludge	Solid	ND	0.059	ND	ND	ND	0.069	0.015	ND	ND	ND
Centrate Recycle Stream	Liquid	< 0.00076	0.0021	< 0.0038	< 0.00076	< 0.00076	< 0.00076	< 0.00076	< 0.00076	< 0.00076	< 0.0038
Centrate Recycle Stream (Duplicate)	Liquid	< 0.00076	0.0019	< 0.0038	< 0.00076	< 0.00076	< 0.00076	< 0.00076	< 0.00076	< 0.00076	< 0.0038
Centrate Recycle Stream	Solid	insufficient solids to centrifuge									

Notes: E = Estimated, ND = Not Detected, NA = Not Applicable, U-D = Result not reported because of sample interferences, D-R = Value is not reported because sample or analyte was ruined. Often due to matrix issues., Q-D = Result not reported because of result did not meet quality criteria, < = Denotes that the analyte was not detected; the associated parameter value is generally the reporting limit, * = Sample was not analyzed for this compound

Table 3-2. Plant A: Instantaneous Loads Results (g/day), Hormones (continued).

Sample Location	Matrix	11-ketotestosterone	19-norethindrone	mestranol	equilenin	17-alpha-EE2	estriol	progesterone	coprostanol (ug/g or ug/L)	cholesterol (ug/g or ug/L)
March 2006										
Thickened Sludge (Combined Primary and Secondary)	Solid	ND	ND	ND	ND	ND	ND	E 0.055	E 0.42	E 1.6
Aerobically Digested Sludge	Solid	E 0.031	ND	ND	ND	ND	ND	ND	E 260	E 280
Dewatered Sludge	Solid	ND	ND	ND	ND	ND	ND	E 0.12	E 1	E 1.8
Centrate Recycle Stream	Liquid	< 0.00044	< 0.00044	< 0.00044	< 0.0011	< 0.00044	< 0.00044	< 0.0022	E 430	E 460
Centrate Recycle Stream	Solid	ND	ND	ND	ND	ND	0.00089	0.0064	E 180	E 37
July 2006										
Thickened Sludge (Combined Primary and Secondary)	Solid	0.034	ND	ND	ND	ND	ND	ND	510	850
Aerobically Digested Sludge	Solid	ND	ND	ND	ND	ND	ND	ND	E 2,800	E 3,100
Dewatered Sludge	Solid	0.16	ND	ND	ND	ND	ND	ND	E 860	E 800
Centrate Recycle Stream	Liquid	D-R								
Centrate Recycle Stream	Solid	insufficient solids to centrifuge								
October 2006										
Thickened Sludge (Combined Primary and Secondary)	Solid	ND	ND	ND	ND	ND	0.13	ND	E 3,200	E 4,900
Thickened Sludge (Combined Primary and Secondary) (Duplicate)	Solid	ND	ND	ND	ND	ND	D-R	ND	E 2,500	E 2,400
Aerobically Digested Sludge	Solid	ND	ND	ND	ND	ND	ND	ND	E 3,400	E 2,200
Dewatered Sludge	Solid	ND	ND	ND	ND	ND	ND	ND	E 3,000	E 1,900
Centrate Recycle Stream	Liquid	< 0.0013	< 0.0013	< 0.0013	< 0.0032	< 0.0013	< 0.0013	< 0.0064	E 39	22
Centrate Recycle Stream	Solid	insufficient solids to centrifuge								
January 2007										
Thickened Sludge (Combined Primary and Secondary)	Solid	ND	ND	ND	ND	ND	ND	E 0.22	E 1.5	E 2.6
Aerobically Digested Sludge	Solid	ND	ND	ND	ND	ND	ND	E 0.19	E 1.4	E 2.5
Dewatered Sludge	Solid	ND	ND	ND	ND	ND	ND	E 0.14	E 0.89	E 1.4
Centrate Recycle Stream	Liquid	D-R	< 0.00076	< 0.00076	< 0.0019	< 0.00076	< 0.00076	< 0.0038	7.2	E 44
Centrate Recycle Stream (Duplicate)	Liquid	D-R	< 0.00076	< 0.00076	< 0.0019	< 0.00076	< 0.00076	< 0.0038	1.6	E 19
Centrate Recycle Stream	Solid	insufficient solids to centrifuge								

Notes: E = Estimated, ND = Not Detected, NA = Not Applicable, U-D = Result not reported because of sample interferences, D-R = Value is not reported because sample or analyte was ruined. Often due to matrix issues., Q-D = Result not reported because of result did not meet quality criteria, < = Denotes that the analyte was not detected; the associated parameter value is generally the reporting limit, * = Sample was not analyzed for this compound

Table 3-3. Plant A: Instantaneous Loads Results (g/day), Alkylphenolic Compounds.

Sample Location	Matrix	4-Nonylphenol		4-nonylphenol monoethoxylates		4-Nonylphenol diethoxylates		4-tert-Octylphenol		4-Octylphenol monoethoxylates		4-Octylphenol diethoxylates	
March 2006													
Aerobically Digested Sludge	Solid	*		*		*		*		*		*	
Dewatered Sludge	Solid	*		*		*		*		*		*	
Centrate Recycle Stream	Liquid	<	0.55	<	0.22	E	0.97	<	0.11	<	0.11	<	0.11
July 2006													
Thickened Sludge (Combined Primary and Secondary)	Solid	E	200	E	930	E	1,300	<	67	E	75	E	53
Aerobically Digested Sludge	Solid	E	270	E	350	E	740	<	100	E	29	<	100
Dewatered Sludge	Solid	E	270	E	530	E	690	E	19	E	41	<	45
Centrate Recycle Stream	Liquid	E	1.4	<	0.44	<	11	E	0.058	<	2.2	<	2.2
October 2006													
Thickened Sludge (Combined Primary and Secondary)	Solid	<	1,800	E	360	E	1,000	<	120	<	600	<	120
Aerobically Digested Sludge	Solid	<	1,100	E	550	E	690	<	71	<	360	<	71
Dewatered Sludge	Solid	E	170	E	370	E	460	E	8.7	E	19	<	17
Centrate Recycle Stream	Liquid	<	0.64	<	0.64	E	2.0	<	0.64	E	0.18	<	0.64
January 2007													
Thickened Sludge (Combined Primary and Secondary)	Solid	<	480	E	320	E	600	<	32	<	160	<	32
Aerobically Digested Sludge	Solid	<	200	E	210	E	360	<	13	<	67	<	13
Dewatered Sludge	Solid	<	300	E	420	E	590	<	20	<	99	<	20
Centrate Recycle Stream	Liquid	<	3.4	E	0.46	<	9.5	E	0.042	<	1.9	<	1.9

Notes: E = Estimated, ND = Not Detected, NA = Not Applicable, U-D = Result not reported because of sample interferences, Q-D = Result not reported because of result did not meet quality criteria, < = Denotes that the analyte was not detected; the associated parameter value is generally the reporting limit, * = Sample was not analyzed for this compound

Table 3-3. Plant A: Instantaneous Loads Results (g/day), Alkylphenolic Compounds (continued).

Sample Location	Matrix	5-Methyl-1H-benzotriazole	Bisphenol A	beta-Sitosterol	1,4-Dichlorobenzene	4-Cumylphenol
March 2006						
Aerobically Digested Sludge	Solid	*	*	*	*	*
Dewatered Sludge	Solid	*	*	*	*	*
Centrate Recycle Stream	Liquid	U-D	E 0.065	< 0.22	< 0.055	< 0.11
July 2006						
Thickened Sludge (Combined Primary and Secondary)	Solid	*	U-D	E 1,100	< 67	< 67
Aerobically Digested Sludge	Solid	*	U-D	E 340	< 100	< 100
Dewatered Sludge	Solid	*	U-D	E 540	< 45	< 45
Centrate Recycle Stream	Liquid	< 4.4	E 0.13	< 4.4	< 1.1	< 2.2
October 2006						
Thickened Sludge (Combined Primary and Secondary)	Solid	*	U-D	E 780	< 120	< 120
Aerobically Digested Sludge	Solid	*	U-D	E 620	< 71	< 71
Dewatered Sludge	Solid	*	E 9.7	E 550	< 17	< 17
Centrate Recycle Stream	Liquid	< 0.64	U-D	< 2.6	< 0.64	< 0.64
January 2007						
Thickened Sludge (Combined Primary and Secondary)	Solid	*	U-D	E 970	< 32	< 32
Aerobically Digested Sludge	Solid	*	U-D	E 380	< 13	< 13
Dewatered Sludge	Solid	*	U-D	E 610	< 20	< 20
Centrate Recycle Stream	Liquid	< 3.4	< 0.76	< 3.8	< 0.15	< 0.27

Notes: E = Estimated, ND = Not Detected, NA = Not Applicable, U-D = Result not reported because of sample interferences, Q-D = Result not reported because of result did not meet quality criteria, < = Denotes that the analyte was not detected; the associated parameter value is generally the reporting limit, * = Sample was not analyzed for this compound

Table 3-4. Plant A: Instantaneous Loads Results, YES Bioassay.

Sample	Eluent Fraction	Instantaneous Load (g/day, EE2 Eqs)			
		Mar-06	Jul-06	Oct-06	Jan-07
Thickened Sludge (Primary & Secondary)	20 50 80	0.0046	1.3599	0.0082	0.0013
Aerobically Digested Sludge	20 50 80	0.0038	1.1154	0.0037	NR
Dewatered Sludge	20 50 80	0.0008	NR	0.0109	0.0005
Centrate Recycle Stream (L)	20 50 80	0.0007	NR/T	0.0037	0.0015
Centrate Recycle Stream (S)	20 50 80	0.0002	M	0.0001	0.0028

Notes: NR = no estrogenic response from sample, T = sample contained toxicity (no estrogenic response was observed), T* = sample contained toxicity (estrogenic response was also observed but was not quantified due to presence of toxicity), M = missing results (samples were received, but were not analyzed)

3.3.2 Chemical Analysis: Data Reduction Results and Discussion

3.3.2.1 Steroids

For Plant A, a comprehensive investigation of all liquid and solid unit processes was not conducted. Rather, there was specific interest in evaluating the efficiency of aerobic digestion. As such, only two unit processes were evaluated: the digestion process and the dewatering process. Three of the potent hormone compounds were detected in this plant, the primary human estrogen (E2) and two of its metabolites (E1, E3).

In March 2006, E1 and E2 are detected in the thickened sludge (feed to the aerobic digester). After digestion E2 is not detected, but E1 load increases by about 25%. This is not surprising because E1 is a known intermediate in aerobic biotransformation of E2. Indeed, the load of E1+E2 decreases slightly, indicating an overall decrease. Although E1, E2, and E3 are observed in the centrate which is recycled back through the plant, none is observed in the dewatered sludge, indicating some transformation has occurred. Approximately 25% of the initial E1 is recycled versus < 5% of the E2, this is consistent with the fact that E1 is an intermediate metabolite of E2 under aerobic conditions. The presence of E3 in the centrate is likely due to concentrations that may have been just below detection levels in other samples.

In July 2006, neither E2 nor E3 was detected. In the absence of E2 as an E1 precursor, E1 is not detected after digestion, indicating likely biotransformation of the incoming E1 to unknown products. However, E1 is detected in the dewatered sludge and its load is substantially higher than in the thickened sludge.

In October 2006, all three estrogens are detected in the thickened sludge. Although a duplicate was analyzed, there were QC failures and the numbers in the primary sample were deemed more reliable. After digestion, E3 is no longer detectable, and both E1 and E2 increase substantially. It is likely that E2 can be produced from E3 by a cleavage of the hydroxyl group in the C16 position on estriol, which could account for an increase in E2 through digestion. This has not been previously demonstrated, to our knowledge, and is merely a hypothesized pathway based on the incoming loads. Still, the total load of E1+E2+E3 increases twofold. Many sources have suggested that cleavage of sulfate- and glucoronide-conjugated hormones is a potential source of free hormones in WWTPs. However, that is unlikely here because the polar conjugates are fairly labile during activated sludge treatment and even if they survived secondary treatment are unlikely to partition into the solid phase in any appreciable amount. More likely, the combination of temporal variation and long SRTs of unit processes make direct comparison of loads difficult. Even though a direct comparison is difficult, it is clear that no solids process in use at this plant has the effect of substantial estrogen degradation.

In January 2007, only E1 was detected at Plant A, and only in the aerobically digested sludge and the dewatered sludge. For the first time, we did not observe E1 in the incoming thickened sludge. Its generation is likely due to previously discussed mechanisms, and its load appears to decrease through dewatering, however there appears to be no source term. Again, this is most likely due to long SRTs with temporal variation of hormone concentrations, in addition to the interpretive difficulties associated with working near analytical detection limits.

When data for the four sampling events are averaged, a somewhat clearer picture emerges. Average load of E1 in the aerobically digested sludge is approximately twice that of the feed to the reactor (thickened sludge), consistent with the documented production of E1 from E2 and likely from E3 as well. The load remains approximately constant through dewatering. The total flux out of the dewatering process is the sum of the load as dewatered sludge and the load in the centrate recycle stream. For E2, the mean concentration in the digested sludge also increases relative to the thickened sludge, likely due to degradation of E3. In contrast to E1, E2 is not detectable in the dewatered sludge. There is likely continued biological activity degrading both E1 and E2 during dewatering. If we assume the following hypothesized reaction scheme under aerobic conditions and the steps have similar rates:

E3 → E2 → E1 → other products

then as long as there is E3 remaining in the system there is unlikely to be any substantial decrease in E2 load. Likewise, as long as there is E2 remaining in the system, there is unlikely to be any substantial decrease in E1 load. So E3 should be the first estrogen to fall below detection levels, followed by E2, followed by E1, consistent with our observations, both for individual sampling events as well as when the four sampling events are averaged. Furthermore, the total load of estrogens (E1+E2+E3) is decreased only slightly through digestion and dewatering. That said, the relative proportion shift away is from E2 in favor of E1, which has less potency. Thus, aerobic digestion may indeed have the effect of reducing steroid-derived estrogenicity. The ramifications of this shift will be discussed further in the context of the YES bioassay results.

3.3.2.2 Non-Estrogenic Steroids

As a rule, the non-estrogenic steroids are more effectively removed by the plant than the estrogenic steroids. This is likely the combination of two factors. First, the lack of an aromatic ring makes them more susceptible to biological processes than the estrogens. Aerobic digestion removes more than 20% of the three androgens present in the digester feed, and after dewatering is complete, more than 40% was removed (Table 3-2). These removals are less than for E2 and E3, which are more recalcitrant during activated sludge treatment. Although liquid samples were not collected at this plant, data from Plants B and D (Tables 3-5 and 3-12, respectively) indicate that androgens typically are present in plant influents at much higher levels than the estrogens, and are removed during activated sludge treatment with greater than 95% effectiveness. Therefore, what remains could be a more refractory segment of the initial input explaining the less than complete removals observed.

3.3.3 Biological Analysis: Data Reduction Results, Model of Concentration Addition, and Discussion

The average (n = 4) mass flux of estrogenicity in and out of the aerobic digester at Plant A was calculated using the Model of Concentration Addition (Figure 3-1), which showed an average estrogenicity flux increase of 450% across the digester. The average estrogenic mass flux, based on YES bioassay measurements, showed a 51% reduction across the digester (Figure 3-2). The estrogenic mass fluxes (based on the YES bioassay measurements) for the pre- and post-digested solids represented 8% and 1%, respectively, of the mass fluxes determined using the Model of Concentration Addition. This disparity in the result between the Concentration Addition Model and the bioassay was observed at other facilities and is discussed further in this section.

Considering the hormones as a group, there was an average increase of 404% through aerobic digestion, attributed mainly to increases in 17β-estradiol (E2) and estrone (E1) (22% and 9%, respectively, of total estrogenicity flux in digested solids) (Figure 3-3). Fluxes of the natural estrogens E1 and E2 increased over 1100% through the digester. Increases in estrone during digestion could be due to deconjugation (D'Ascenzo, 2003) and/or aerobic degradation of 17β-estradiol to form estrone as a metabolite (Scherr, 2009; Colucci, 2001; Lee, 2003; Ying, 2005). Estrone increased through the digester for three of the four sampling periods. 17β-estradiol was not detected in three of the four thickened combined sludge samples and also not detected in the July 2006 and January 2007 digested solid samples. Estriol (E3) was detected in only one out of four sampling dates, October 2006, and was only detected in the thickened combined sludge (and at <1% of total estrogenicity mass flux).

The alkylphenol group (APEOs) exhibited a 17% increase across the aerobic digester (Figure 3-4), DEHP decreased by 43%, and DES increased 598% across the digester (Figure 3-5). DES was a more important contributor to estrogenic mass flux at Plant A than at any of the other Plants in the study and comprised the majority of the total estrogenicity in the aerobically digested solids at Plant A, increasing by about one order of magnitude. As discussed in sub-section 3.2.1, due to improvements in GC/MS/MS analysis of DES over the course of the study, and the lack of a likely major source term for DES, the confidence in this conclusion is less than for the other hormones and estrogenic AWIs.

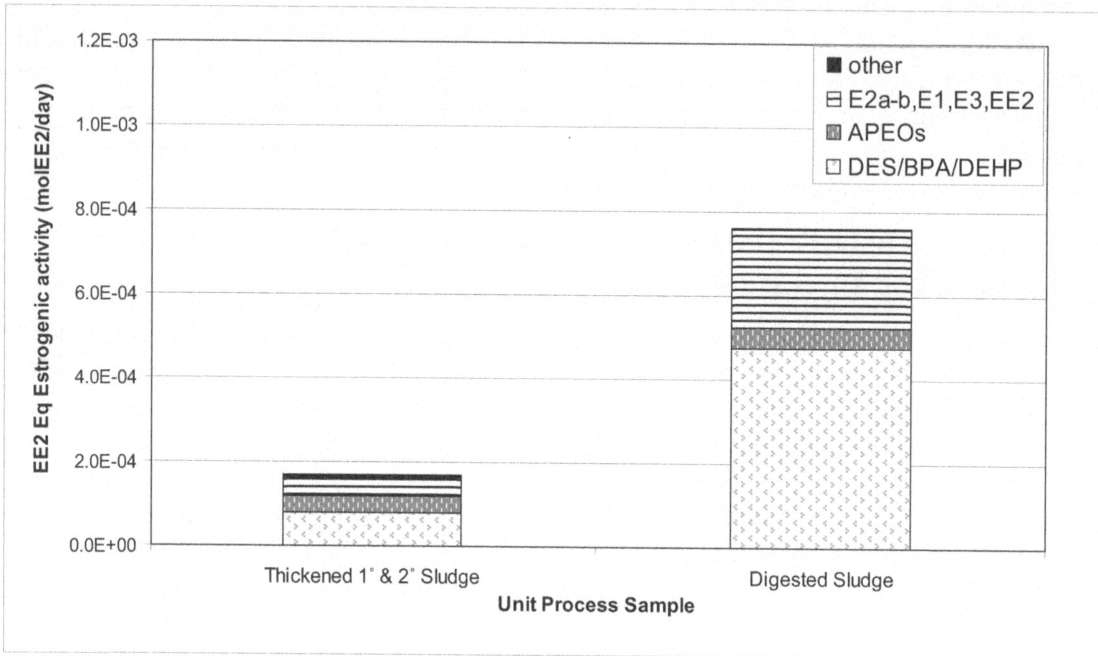

Figure 3-1. Average Daily Mass Fluxes (mol EE2-equivalents/day) of Estrogenicity Before and After Aerobic Digestion at Plant A. (Based on the Model of Concentration Addition)

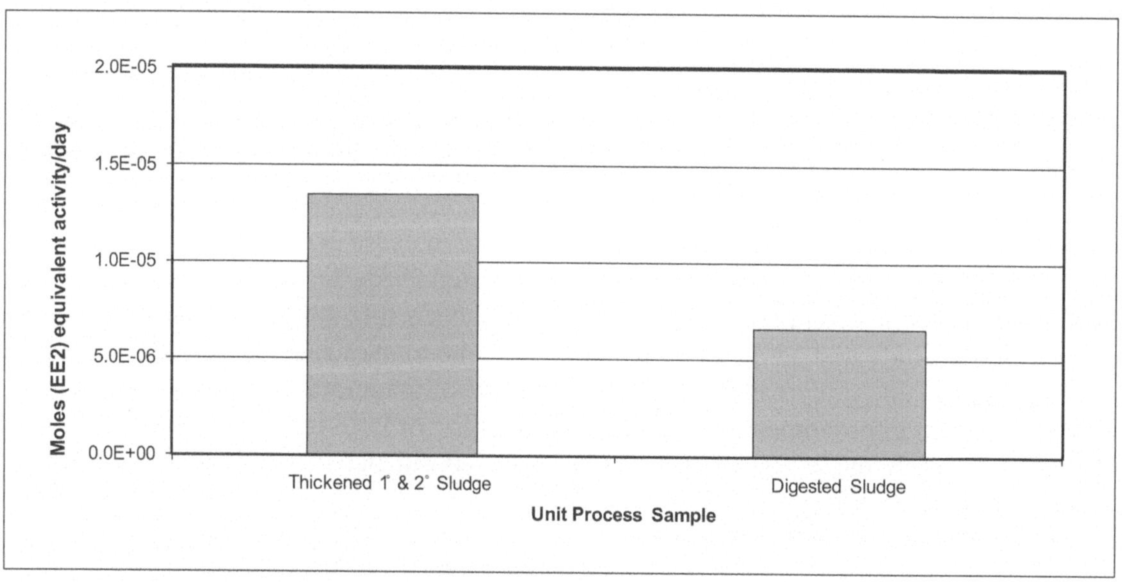

Figure 3-2. Average Daily Mass Fluxes (mol EE2-equivalents/day) of Estrogenic Activity Before and After Aerobic Digestion at Plant A. (Based on YES Bioassay Measurements)

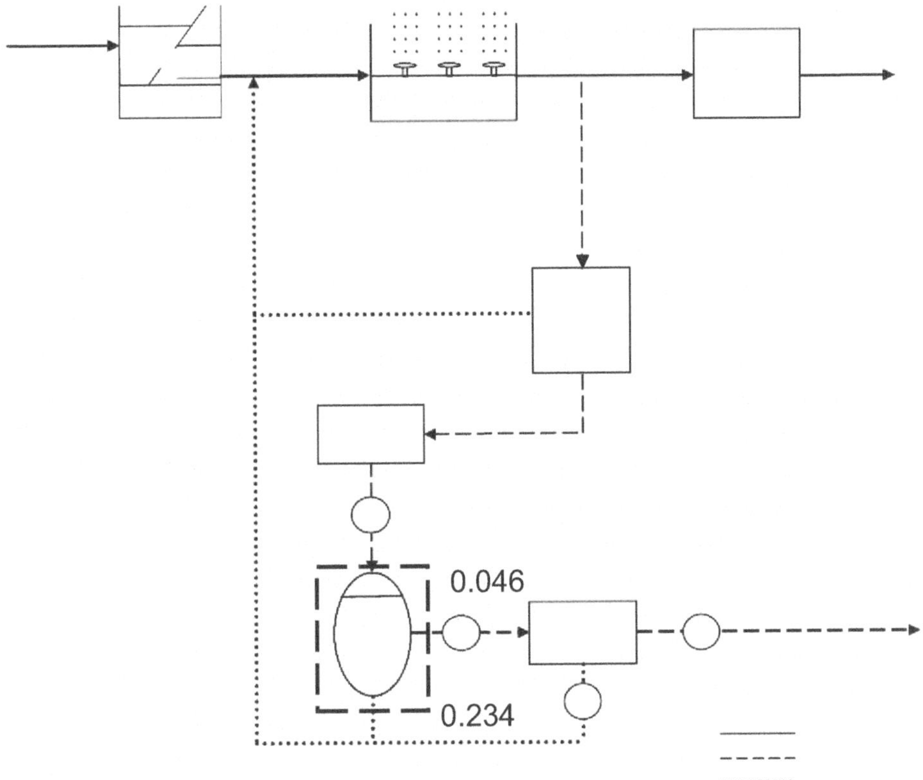

Figure 3-3. Daily Estrogenicity Mass Flux (mmol EE2 equivalents/day) due to Estrogenic Hormones (E2-α,E2-β,E1,E3, EE2) at Plant A. (Based on the Model of Concentration Addition)

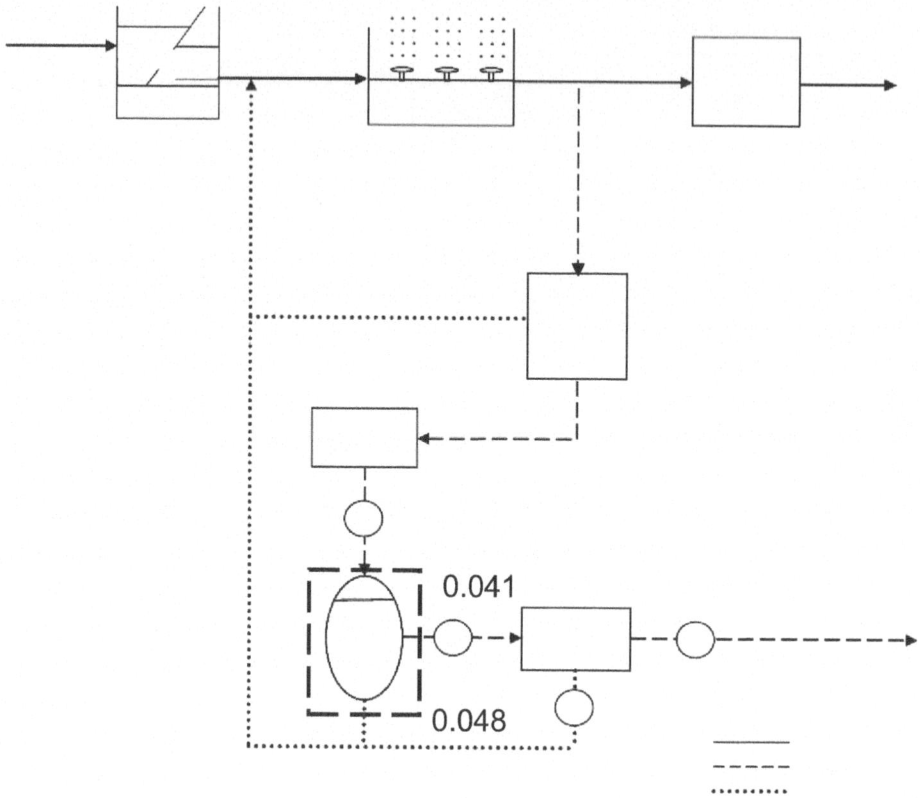

Figure 3-4. Daily Estrogenicity Mass Flux (mmol EE2 equivalents/day) Provided by Total APEOs at Plant A. (Based on the Model of Concentration Addition)

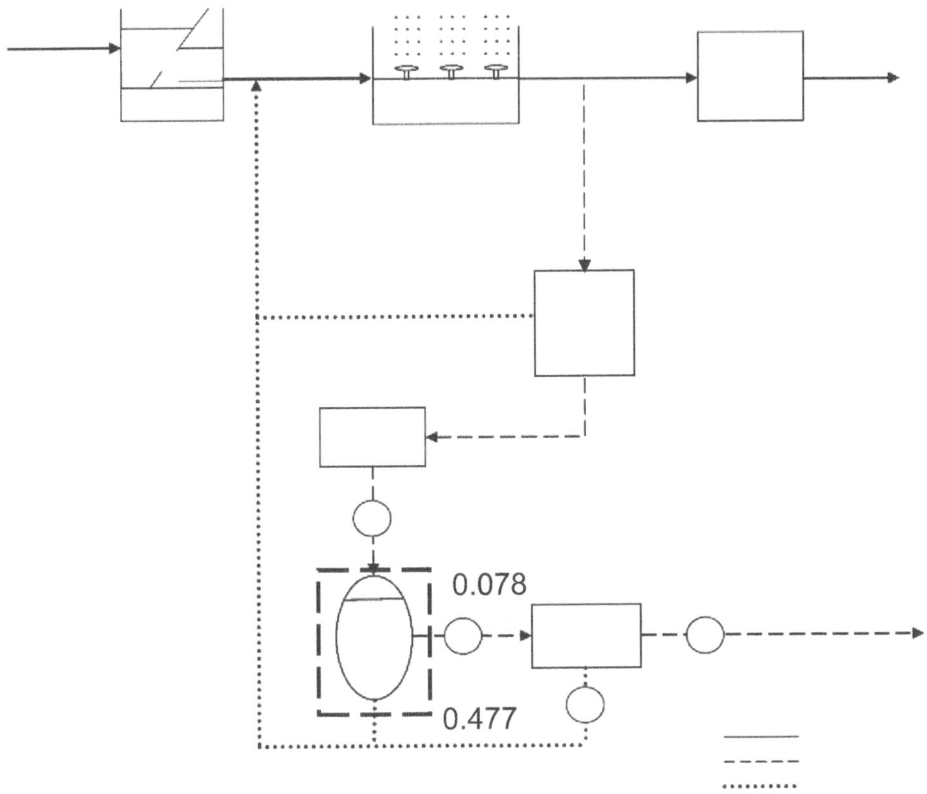

Figure 3-5. Daily Mass Flux (mmol EE2 equivalents/day) of DES/BPA/DEHP at Plant A. (Based on the Model of Concentration Addition)

3.4 Plant B

3.4.1 Instantaneous Load: Hormones, Alkylphenolic Compounds, and Bioassays

Instantaneous loads calculations for all sample dates for steroid hormones and alkylphenolic compounds are provided in Tables 3-5 and 3-6, respectively. Table 3-7 shows the instantaneous loads results for the YES bioassay at Plant B.

It was difficult to close the instantaneous loads balance based on the different types of digestion processes operated at Plant B since samples were only collected from the conventional anaerobic digesters in December 2005. Additionally, the dual phase digester and the egg-shaped anaerobic digesters received approximately three quarters of the primary sludge and roughly half of the thickened waste activated sludge (TWAS) flow while the conventional anaerobic digesters received approximately one quarter of the primary sludge flow and the other half of the TWAS flow. For this sample period, the loads from all three digestion processes were similar. The loads of target hormones were lowest from the dual phase digested sludge and highest from the conventional digesters. Detected concentrations of the target hormones were actually lowest coming out of the egg-shaped digesters. There was a decrease in the load of target steroid hormones from the acid to the methane phase in the dual phase digester. For all sample dates, concentrations of estriol increased in dewatered sludges compared to digested sludge samples.

All target alkylphenolic compounds persisted in dewatered and pelletized sludge. Although not all digested sludge was accounted for in each sample period, using the conservative estimate that all digestion processes had similar loads of target hormones it appears that the loads of several increased in the dewatering step for all sample periods. Further, although not all of the dewatered sludge was pelletized, these samples showed decreased loads compared to the dewatered sludge. These observations correspond with the loads of estrogenicity observed between the digested sludges, dewatered sludge and the pelletized sludge.

For the one date that the composted sludge was analyzed, the loads of hormones were comparable to that of the pelletized sludge; however, there is no data for the loads of alkylphenolic compounds in the composted sludge sample. The load of estrogenicity in the composted sludge was higher than that in the pelletized sludge (0.0773 vs. 0.0073 g/day EE2 Eqs).

There is a complete dataset for liquid streams in January 2007. During this period, most target hormones were detected in the plant influent. The majority of loads were reduced to non-detect following secondary treatment. Although there is a lack of data on the loads of alkylphenolic compounds in the liquid streams, loads in the solids stream indicate that the loads of these analytes are not substantially reduced in the dewatered sludge, as stated above. For dates when liquid streams were sampled, the bioassay results show a reduction in estrogenic activity following secondary treatment in both the liquid and solid streams. Lower loads of estrogenic activity were present in centrate compared to dewatered sludge.

Table 3-5. Plant B: Instantaneous Loads Results (g/day), Hormones.

Sample Location	Matrix	diethylstilbestrol	cis-androsterone	epitestosterone	17-alpha-estradiol	dihydrotestosterone	androstene-3,17-dione	estrone	17-beta-estradiol	testosterone	equilin
December 2005											
Thickened Sludge (Combined Primary & Secondary)	Solid	ND	6.6	ND	ND	1.6	7.6	2.6	ND	0.70	ND
Anaerobically Digested Sludge	Solid	ND	E 7.2	ND	ND	ND	E 2.2	1.8	1.8	E 0.075	ND
Acid Phase Digested Sludge	Solid	ND	E 22	ND	ND	E 0.63	E 1.3	E 1.7	ND	E 0.29	ND
Methane Phase Digested Sludge	Solid	ND	E 6.3	ND	ND	E 0.13	E 2.8	E 0.51	ND	E 0.053	ND
Conventional Digested Sludge	Solid	ND	E 9.2	ND	ND	E 0.56	E 7.6	E 3.9	ND	ND	ND
Dewatered Sludge (Pelletech)	Solid	ND	E 16	ND	ND	ND	E 30	E 290	ND	E 1.5	ND
Dewatered Sludge (Pelletech) (Duplicate, extracted 9/18)	Solid	0.40	20	ND	0.16	1.5	49	29	0.57	5.9	ND
Dewatered Sludge (Pelletech) (Duplicate)	Solid	ND	2.0	ND	ND	ND	12	1.7	ND	ND	ND
Centrate Recycle Stream (Pelletech) (solid)	Solid	D-R	D-R	D-R	D-R	D-R	D-R	D-R	D-R	D-R	D-R
Tertiary Pelletized Sludge (Pelletech)	Solid	0.090	28	ND	0.34	0.57	0.98	2.0	0.31	ND	ND
February 2006											
Centrate Recycle Stream (Pelletech) (liquid)*	Liquid	< 0.059	E 83	< 0.30	1.8	2.6	39	25	1.1	1.8	< 0.30
March 2006											
Thickened Sludge (Combined Primary & Secondary)	Solid	ND	E 9.1	E 1.8	E 0.17	E 1.2	E 5.9	E 2.7	ND	E 2.6	ND
Anaerobically Digested Sludge	Solid	ND	E 4.5	ND	E 0.089	E 0.28	E 5.1	E 1.5	ND	E 0.096	ND
Dewatered Sludge (Pelletech)	Solid	ND	E 200	ND	ND	E 3.3	E 48	E 29	ND	E 1.5	ND
Centrate Recycle Stream (Pelletech) (solid)	Solid	ND	1.8	ND	0.0094	0.16	0.30	0.36	0.099	0.050	ND
Centrate Recycle Stream (Pelletech) (solid) (Duplicate)	Solid	ND	0.88	ND	ND	0.027	0.38	0.29	ND	0.014	ND
Centrate Recycle Stream (Pelletech) (solid) (Triplicate)	Solid	ND	1.3	ND	ND	0.050	0.65	0.35	ND	0.024	ND
July 2006											
Primary Influent	Liquid	D-R	D-R	D-R	D-R	D-R	D-R	D-R	D-R	D-R	D-R
Primary Effluent	Liquid	< 0.46	E 4,900	< 2.3	< 0.46	150	480	57	18	63	< 2.3
Secondary Effluent	Liquid	< 0.46	< 0.46	< 2.3	< 0.46	< 0.46	< 0.46	< 0.46	< 0.46	< 0.46	< 2.3
Primary Unthickened Sludge	Solid	ND	ND	ND	ND	9.6	D-R	D-R	ND	ND	ND
Secondary Unthickened Sludge	Solid	D-R	D-R	D-R	D-R	D-R	D-R	D-R	D-R	D-R	D-R
Anaerobically Digested Sludge	Solid	ND	2.9	ND	0.075	ND	4.5	0.91	0.048	ND	ND
Acid Phase Digested Sludge	Solid	ND	0.87	ND	ND	1.2	3.2	0.58	0.066	0.36	E 0.20
Methane Phase Digested Sludge	Solid	ND	E 4.5	ND	0.026	0.53	0.71	0.37	0.086	0.099	ND
Dewatered Sludge (Pelletech)	Solid	ND	4.7	ND	0.13	ND	11	5.1	ND	0.26	ND
Centrate Recycle Stream (Pelletech) (liquid)	Liquid	< 0.059	E 7.4	< 0.30	0.17	0.72	2.1	1.8	0.10	0.11	< 0.30
Centrate Recycle Stream (Pelletech) (solid)	Solid	ND	0.92	ND	0.0058	0.044	0.19	0.026	ND	ND	ND
Tertiary Pelletized Sludge (Pelletech)	Solid	ND	E 10	ND	0.073	0.42	1.2	2.2	0.11	0.13	ND

Notes: E = Estimated, ND = Not Detected, NA = Not Applicable, U-D = Result not reported because of sample interferences, D-R = Value is not reported because sample or analyte was ruined. Often due to matrix issues., Q-D = Result not reported because of result did not meet quality criteria, < = Denotes that the analyte was not detected; the associated parameter value is generally the reporting limit, * = Sample was not analyzed for this compound

Table 3-5. Plant B: Instantaneous Loads Results (g/day), Hormones (continued).

Sample Location	Matrix	11-ketotestosterone	19-norethindrone	mestranol	equilenin	17-alpha-EE2	estriol	progesterone	coprostanol	cholesterol
December 2005										
Thickened Sludge (Combined Primary & Secondary)	Solid	ND	ND	ND	ND	ND	D-R	E 9.9	E 110	E 130
Anaerobically Digested Sludge	Solid	E 0.051	ND	ND	ND	ND	ND	E 3.8	E 9.9	E 3.2
Acid Phase Digested Sludge	Solid	E 0.31	ND	ND	ND	ND	ND	E 1.1	E 6.2	E 6.9
Methane Phase Digested Sludge	Solid	ND	ND	ND	ND	ND	ND	E 6.2	E 2.7	E 1.3
Conventional Digested Sludge	Solid	E 0.39	ND	ND	ND	ND	ND	E 5.3	E 7.8	E 2.0
Dewatered Sludge (Pelletech)	Solid	E 1.2	ND	ND	E	ND	ND	E 8.5	E 22	E 7.1
Dewatered Sludge (Pelletech) (Duplicate, extracted 9/18)	Solid	ND	ND	ND	ND	ND	ND	E 20	E 300	E 130
Dewatered Sludge (Pelletech) (Duplicate)	Solid	ND	ND	ND	ND	ND	D-R	E 7.1	E 33	E 13
Centrate Recycle Stream (Pelletech) (solid)	Solid	D-R	D-R	D-R	D-R	D-R	D-R	D-R	D-R	D-R
Tertiary Pelletized Sludge (Pelletech)	Solid	0.48	ND	ND	ND	ND	ND	ND	E	E
February 2006										
Centrate Recycle Stream (Pelletech) (liquid)*	Liquid	6.0	< 0.059	< 0.059	0.62	< 0.059	3.7	E 16	1.6	0.83
March 2006										
Thickened Sludge (Combined Primary & Secondary)	Solid	E 0.51	ND	ND	ND	ND	ND	E 4.6	E 26	E 27
Anaerobically Digested Sludge	Solid	ND	ND	ND	ND	ND	ND	E 12	E 9.9	E 3.4
Dewatered Sludge (Pelletech)	Solid	E 0.30	ND	ND	ND	ND	ND	E 65	E 57	E 20
Centrate Recycle Stream (Pelletech) (solid)	Solid	ND	ND	ND	ND	ND	ND	E 0.38	E 11	E 4.6
Centrate Recycle Stream (Pelletech) (solid) (Duplicate)	Solid	ND	ND	ND	ND	ND	ND	E 2.3	E 3.1	E 1.1
Centrate Recycle Stream (Pelletech) (solid) (Triplicate)	Solid	0.022	ND	ND	ND	ND	ND	E 2.8	E 3.8	E 1.3
July 2006										
Primary Influent	Liquid	D-R	D-R	D-R	D-R	D-R	D-R	D-R	D-R	D-R
Primary Effluent	Liquid	50	< 0.46	< 0.46	< 1.1	< 0.46	300	E 130	E 490	E 630
Secondary Effluent	Liquid	< 0.46	< 0.46	< 0.46	< 1.2	< 0.46	< 0.46	< 2.3	< 2.3	< 2.3
Primary Unthickened Sludge	Solid	ND	ND	ND	ND	ND	D-R	ND	E	E
Secondary Unthickened Sludge	Solid	D-R	D-R	D-R	D-R	D-R	D-R	D-R	D-R	D-R
Anaerobically Digested Sludge	Solid	ND	ND	ND	ND	ND	ND	E 3.4	E 13	E 6.5
Acid Phase Digested Sludge	Solid	ND	ND	ND	ND	ND	ND	E 11	E 5.4	E 7.6
Methane Phase Digested Sludge	Solid	ND	ND	ND	ND	ND	ND	E 1.1	E 8.9	E 5.4
Dewatered Sludge (Pelletech)	Solid	ND	ND	ND	ND	ND	ND	E 23	E 58	E 39
Centrate Recycle Stream (Pelletech) (liquid)	Liquid	0.25	< 0.059	< 0.059	< 0.15	< 0.059	0.0 1.4	E 2.4	2.4	1.0
Centrate Recycle Stream (Pelletech) (solid)	Solid	ND	ND	ND	ND	ND	ND	E 0.094	E 7.4	E 9.7
Tertiary Pelletized Sludge (Pelletech)	Solid	ND	ND	ND	ND	ND	ND	E 1.8	E 15	E 9.4

Notes: E = Estimated, ND = Not Detected, NA = Not Applicable, U-D = Result not reported because of sample interferences, D-R = Value is not reported because sample or analyte was ruined. Often due to matrix issues., Q-D = Result not reported because of result did not meet quality criteria, < = Denotes that the analyte was not detected; the associated parameter value is generally the reporting limit, * = Sample was not analyzed for this compound

Table 3-5. Plant B: Instantaneous Loads Results (g/day), Hormones (continued).

Sample Location	Matrix	diethylstilbestrol	cis-androsterone	epitestosterone	17-alpha-estradiol	dihydrotestosterone	androstene-3,17-dione	estrone	17-beta-estradiol	testosterone	equilin
October 2006											
Thickened Sludge (Combined Primary & Secondary)	Solid	ND	24	ND	ND	ND	7.4	2.9	ND	ND	ND
Anerobically Digested Sludge	Solid	ND	2.9	ND	0.18	ND	4.8	2.7	ND	ND	ND
Dewatered Sludge (Pelletech)	Solid	ND	14	ND	0.28	0.99	6.6	4.1	ND	ND	ND
Centrate Recycle Stream (Pelletech) (liquid)	Liquid	0.013	< 0.059	< 0.30	< 0.059	< 0.059	1.1	0.13	< 0.059	< 0.059	< 0.30
January/February 2007											
Primary Influent	Liquid	< 0.47	E 2,600	< 2.3	< 0.47	94	82	14	4.3	24	< 2.3
Primary Effluent	Liquid	< 0.46	E 1,800	< 2.3	< 0.46	74	74	14	4.2	23	< 2.3
Primary Effluent (Duplicate)	Liquid	< 0.46	E 1,700	< 2.3	< 0.46	74	69	14	3.0	20	< 2.3
Secondary Effluent	Liquid	< 0.46	0.40	< 2.3	< 0.46	< 0.46	2.8	22	3.3	< 0.46	< 2.3
Secondary Effluent (Duplicate)	Liquid	< 0.46	1.1	< 2.3	< 0.46	< 0.46	3.4	18	1.2	< 0.46	< 2.3
Primary Unthickened Sludge	Solid	ND	38	ND	ND	4.1	4.5	1.9	ND	ND	ND
Secondary Unthickened Sludge	Solid	D-R	D-R	D-R	D-R	D-R	D-R	D-R	D-R	D-R	D-R
Anerobically Digested Sludge	Solid	D-R	1.9	ND	ND	ND	2.1	1.9	ND	ND	ND
Anerobically Digested Sludge (Duplicate)	Solid	D-R	2.6	ND	ND	ND	2.2	2.0	ND	ND	ND
Acid Phase Digested Sludge	Solid	D-R	5.1	ND	ND	0.86	1.1	0.81	0.20	0.92	ND
Methane Phase Digested Sludge	Solid	D-R	5.1	ND	ND	ND	0.29	0.49	ND	ND	ND
Methane Phase Digested Sludge (Duplicate)	Solid	ND	4.3	ND	ND	0.34	0.32	0.43	0.066	0.13	ND
Dewatered Sludge (Pelletech)	Solid	1.5	12	ND	0.56	0.78	13	6.3	ND	ND	ND
Centrate Recycle Stream (Pelletech) (liquid)	Liquid	< 0.059	E 6.4	< 0.30	0.44	< 0.059	3.1	3.7	< 0.059	< 0.059	< 0.30
Tertiary Pelletized Sludge (Pelletech)	Solid	ND	14	ND	ND	0.69	1.3	2.2	0.30	0.20	ND
Tertiary Pelletized Sludge (Pelletech) (Duplicate)	Solid	0.077	13	ND	0.16	0.53	1.3	2.2	0.31	0.18	ND
Tertiary composted sludge	Solid	0.18	0.80	ND	ND	0.43	4.2	8.8	0.074	0.070	ND
Tertiary composted sludge (Duplicate)	Solid	ND	1.2	ND	ND	ND	4.3	11	ND	ND	ND
May 2007											
Secondary Effluent*	Liquid	< 0.46	0.58	< 2.3	< 0.46	< 0.46	4.6	6.6	0.29	< 0.46	< 2.3
Secondary Effluent (Duplicate)*	Liquid	< 0.46	0.46	< 2.3	< 0.46	< 0.46	4.1	6.7	< 0.46	< 0.46	< 2.3
Secondary Effluent (Triplicate)*	Liquid	< 0.46	0.52	< 2.3	< 0.46	< 0.46	4.4	7.2	0.40	< 0.46	< 2.3

Notes: E = Estimated, ND = Not Detected, NA = Not Applicable, U-D = Result not reported because of sample interferences, D-R = Value is not reported because sample or analyte was ruined. Often due to matrix issues., Q-D = Result not reported because of result did not meet quality criteria, < = Denotes that the analyte was not detected; the associated parameter value is generally the reporting limit, * = Sample was not analyzed for this compound

Table 3-5. Plant B: Instantaneous Loads Results (g/day), Hormones (continued).

Sample Location	Matrix	11-ketotestosterone	19-norethindrone	mestranol	equilenin	17-alpha-EE2	estriol	progesterone	coprostanol	cholesterol
October 2006										
Thickened Sludge (Combined Primary & Secondary)	Solid	1.2	ND	ND	ND	ND	ND	E 16	E 74	E 99
Anerobically Digested Sludge	Solid	0.49	ND	ND	ND	ND	ND	E 27	E 150	E 62
Dewatered Sludge (Pelletech)	Solid	0.35	ND	ND	ND	ND	ND	E 29	E 260	E 92
Centrate Recycle Stream (Pelletech) (liquid)	Liquid	<	< 0.059	0.059	< 0.15	< 0.059	ND	< 0.30	1.3	0.73
January/February 2007										
Primary Influent	Liquid	65	< 0.47	< 0.47	< 1.2	< 0.47	110	E 12	E 270	E 190
Primary Effluent	Liquid	42	< 0.46	< 0.46	< 1.1	< 0.46	91	E 17	E 480	E 310
Primary Effluent (Duplicate)	Liquid	49	< 0.46	< 0.46	< 1.1	< 0.46	86	E 11	E 370	E 280
Secondary Effluent	Liquid	< 0.46	< 0.46	< 0.46	< 1.2	< 0.46	< 0.46	< 2.3	4.7	4.7
Secondary Effluent (Duplicate)	Liquid	< 0.46	< 0.46	< 0.46	< 1.2	< 0.46	< 0.46	< 2.3	20	20
Primary Unthickened Sludge	Solid	1.4	ND	ND	ND	ND	1.4	ND	E	E
Secondary Unthickened Sludge	Solid	D-R	D-R	D-R	D-R	D-R	D-R	D-R	D-R	D-R
Anerobically Digested Sludge	Solid	ND	ND	ND	ND	ND	ND	E 8.9	E 58	E 22
Anerobically Digested Sludge (Duplicate)	Solid	ND	ND	ND	ND	ND	ND	E 8.9	E 86	E 33
Acid Phase Digested Sludge	Solid	ND	ND	ND	ND	ND	D-R	ND	E	E
Methane Phase Digested Sludge	Solid	ND	ND	ND	ND	ND	ND	ND	E	E
Methane Phase Digested Sludge (Duplicate)	Solid	ND	ND	ND	ND	ND	D-R	ND	E	E
Dewatered Sludge (Pelletech)	Solid	ND	ND	ND	ND	ND	ND	E 7.7	E 350	E 120
Centrate Recycle Stream (Pelletech) (liquid)	Liquid	< 0.012	< 0.059	< 0.059	< 0.15	< 0.059	2.2	E 2.1	0.50	0.28
Tertiary Pelletized Sludge (Pelletech)	Solid	ND	ND	ND	ND	ND	0.085	E 2.3	E 130	E 49
Tertiary Pelletized Sludge (Pelletech) (Duplicate)	Solid	ND	ND	ND	ND	ND	ND	E 2.0	E 150	E 49
Tertiary composted sludge	Solid	ND	ND	ND	0.68	ND	ND	E 0.45	E 6.6	E 3.7
Tertiary composted sludge (Duplicate)	Solid	ND	ND	ND	0.75	ND	ND	ND	E	
May 2007										
Secondary Effluent*	Liquid	< 0.46	< 0.46	< 0.46	< 1.2	< 0.46	< 0.46	< 2.3	4.1	6.0
Secondary Effluent (Duplicate)*	Liquid	< 0.46	< 0.46	< 0.46	< 1.2	< 0.46	< 0.46	< 2.3	4.4	5.3
Secondary Effluent (Triplicate)*	Liquid	< 0.46	< 0.46	< 0.46	< 1.2	< 0.46	< 0.46	< 2.3	4.5	6.0

Notes: E = Estimated, ND = Not Detected, NA = Not Applicable, U-D = Result not reported because of sample interferences, D-R = Value is not reported because sample or analyte was ruined. Often due to matrix issues., Q-D = Result not reported because of result did not meet quality criteria, < = Denotes that the analyte was not detected; the associated parameter value is generally the reporting limit, * = Sample was not analyzed for this compound

Table 3-6. Plant B: Instantaneous Loads Results (g/day), Alkylphenolic Compounds.

Sample Location	Matrix	4-Nonylphenol	4-nonylphenol monoethoxylates	4-Nonylphenol diethoxylates	4-tert-Octylphenol	4-Octylphenol monoethoxylates	4-Octylphenol diethoxylates
December 2005							
Centrate Recycle Stream (Pelletech) (for lab work)	Liquid	< 74	< 30	< 7,400	< 1,500	< 1,500	< 1,500
Centrate Recycle Stream (Pelletech) (liquid)	Liquid	E 4,700	< 30	< 7,400	< 1,500	< 1,500	< 1,500
April 2006							
Thickened Sludge (Combined Primary & Secondary)	Solid	E 360	E 690	E 510	E 8.1	E 14	< 19
Anaerobically Digested Sludge	Solid	E 2,400	E 150	E 140	29	< 39	< 7.7
Dewatered Sludge (Pelletech)	Solid	E 11,000	E 550	E 580	120	< 85	< 17
Centrate Recycle Stream (Pelletech) (liquid)	Liquid	E 470	E 68	< 74	< 15	< 15	< 15
July 2006							
Primary Influent	Liquid	E 240,000	E 28,000	E 130,000	E 10,000	E 8,000	< 29,000
Primary Effluent	Liquid	E 9,900	E 15,000	E 6,900	E 450	E 520	< 570
Secondary Effluent	Liquid	< 29,000	< 1,200	< 29,000	E 120	< 5,800	< 5,800
Secondary Effluent	Liquid	< 2,900	< 1,200	< 2,900	< 580	< 580	< 580
Primary Unthickened Sludge	Solid	< 30,000	< 20,000	< 40,000	< 2,000	< 10,000	< 2,000
Secondary Unthickened Sludge	Solid	< 48,000	< 32,000	< 64,000	< 3,200	< 16,000	< 3,200
Anaerobically Digested Sludge	Solid	E 31,000	E 2,600	< 11,000	580	< 2,900	< 570
Acid Phase Digested Sludge	Solid	E 3,100	E 4,500	E 2,200	E 87	E 120	< 130
Methane Phase Digested Sludge	Solid	E 5,800	E 570	< 3,400	E 110	< 850	< 170
Dewatered Sludge (Pelletech)	Solid	E 40,000	E 4,200	< 11,000	750	< 2,800	< 550
Centrate Recycle Stream (Pelletech) (liquid)	Liquid	E 390	< 30	< 740	E 17	< 150	< 150
Tertiary Pelletized Sludge (Pelletech)	Solid	E 18,000	E 2,600	E 1,300	240	E 47	< 56

Notes: E = Estimated, ND = Not Detected, NA = Not Applicable, U-D = Result not reported because of sample interferences, Q-D = Result not reported because of result did not meet quality criteria, < = Denotes that the analyte was not detected; the associated parameter value is generally the reporting limit, * = Sample was not analyzed for this compound

Table 3-6. Plant B: Instantaneous Loads Results (g/day), Alkylphenolic Compounds (continued).

Sample Location	Matrix	5-Methyl-1H-benzotriazole	Bisphenol A	beta-Sitosterol	1,4-Dichlorobenzene	4-Cumylphenol
December 2005						
Centrate Recycle Stream (Pelletech) (for lab work)	Liquid	< 3,000	U-D	< 3,000	< 740	< 1,500
Centrate Recycle Stream (Pelletech) (liquid)	Liquid	< 3,000	U-D	< 3,000	< 740	< 1,500
April 2006						
Thickened Sludge (Combined Primary & Secondary)	Solid	*	E 21	E 1,200	< 19	< 19
Anaerobically Digested Sludge	Solid	*	E 23	E 850	E 1.2	< 7.7
Dewatered Sludge (Pelletech)	Solid	*	E 700	E 4,200	E 4.3	< 17
Centrate Recycle Stream (Pelletech) (liquid)	Liquid	< 30	E 3.6	E 9.0	< 7.4	< 15
July 2006						
Primary Influent	Liquid	< 59,000	E 12,000	E 97,000	E 8,700	< 29,000
Primary Effluent	Liquid	< 1,100	E 100	E 3,400	E 580	< 570
Secondary Effluent	Liquid	E 1,800	E 1,200	< 12,000	E 390	< 5,800
Secondary Effluent	Liquid	< 1,200	U-D	< 1,200	< 290	< 580
Primary Unthickened Sludge	Solid	*	U-D	E 4,100	< 2,000	< 2,000
Secondary Unthickened Sludge	Solid	*	U-D	E 5,000	< 3,200	< 3,200
Anaerobically Digested Sludge	Solid	*	E 190	E 12,000	< 570	< 570
Acid Phase Digested Sludge	Solid	*	U-D	E 3,700	< 130	< 130
Methane Phase Digested Sludge	Solid	*	E 96	E 2,200	< 170	< 170
Dewatered Sludge (Pelletech)	Solid	*	E 870	E 19,000	E 550	< 550
Centrate Recycle Stream (Pelletech) (liquid)	Liquid	< 300	E 15	E 190	E 18	< 150
Tertiary Pelletized Sludge (Pelletech)	Solid	*	E 1,400	E 9,100	< 56	< 56

Notes: E = Estimated, ND = Not Detected, NA = Not Applicable, U-D = Result not reported because of sample interferences, Q-D = Result not reported because of result did not meet quality criteria, < = Denotes that the analyte was not detected; the associated parameter value is generally the reporting limit, * = Sample was not analyzed for this compound

Table 3-6. Plant B: Instantaneous Loads Results (g/day), Alkylphenolic Compounds (continued).

Sample Location	Matrix	4-Nonylphenol	4-nonylphenol monoethoxylates	4-Nonylphenol diethoxylates	4-tert-Octylphenol	4-Octylphenol monoethoxylates	4-Octylphenol diethoxylates
October 2006							
Thickened Sludge (Combined Primary & Secondary)	Solid	E 19,000	E 30,000	E 8,700	E 320	E 530	< 610
Anerobically Digested Sludge	Solid	E 25,000	E 780	E 660	310	< 680	< 140
Dewatered Sludge (Pelletech)	Solid	E 89,000	E 2,200	E 1,500	1,200	< 1,100	< 230
Centrate Recycle Stream (Pelletech) (liquid)	Liquid	< 27	< 30	< 74	15	E 7.8	E 2.6
January 2007							
Primary Influent	Liquid	E 14,000	E 16,000	E 14,000	650	E 1,100	E 1,000
Primary Effluent	Liquid	E 13,000	E 10,000	E 11,000	660	E 980	E 760
Secondary Effluent	Liquid	E 640	E 1300	E 3,900	E 49	E 190	E 370
Primary Unthickened Sludge	Solid	E 1,200,000	E 1,500,000	< 2,900,000	< 150,000	< 730,000	< 150,000
Secondary Unthickened Sludge	Solid	< 18,000	E 6,200	< 24,000	< 1,200	< 6,000	< 1,200
Anerobically Digested Sludge	Solid	E 35,000	E 3,100	E 3,000	640	< 960	< 190
Dewatered Sludge (Pelletech)	Solid	E 180,000	E 11,000	E 11,000	2,800	< 3,000	< 590
Centrate Recycle Stream (Pelletech) (liquid)	Liquid	E 280	E 75	< 12	18	< 47	E 10.0
Tertiary Pelletized Sludge (Pelletech)	Solid	E 19,000	E 1,300	E 1,700	250	< 290	< 58

Notes: E = Estimated, ND = Not Detected, NA = Not Applicable, U-D = Result not reported because of sample interferences, Q-D = Result not reported because of result did not meet quality criteria, < = Denotes that the analyte was not detected; the associated parameter value is generally the reporting limit, * = Sample was not analyzed for this compound

Table 3-6. Plant B: Instantaneous Loads Results (g/day), Alkylphenolic Compounds (continued).

Sample Location	Matrix	5-Methyl-1H-benzotriazole	Bisphenol A	beta-Sitosterol	1,4-Dichlorobenzene	4-Cumylphenol
October 2006						
Thickened Sludge (Combined Primary & Secondary)	Solid	*	E 240	E 12000	< 800	< 800
Anerobically Digested Sludge	Solid	*	E 320	E 13000	< 140	< 140
Dewatered Sludge (Pelletech)	Solid	*	E 920	E 410000	E 230	< 230
Centrate Recycle Stream (Pelletech) (liquid)	Liquid	E 13	U-D	< 30	< 1.2	< 2.1
January 2007						
Primary Influent	Liquid	< 3800	E 190	E 3600	E 570	< 470
Primary Effluent	Liquid	< 3700	U-D	E 2600	E 500	< 460
Secondary Effluent	Liquid	E 350	U-D	E 1600	E 190	< 81
Primary Unthickened Sludge	Solid	*	U-D	E 2000000	< 150000	< 150000
Secondary Unthickened Sludge	Solid	*	U-D	E 13000	< 1200	< 1200
Anerobically Digested Sludge	Solid	*	E 4800	E 20000	< 190	< 190
Dewatered Sludge (Pelletech)	Solid	*	E 1500	E 92000	E 200	< 590
Centrate Recycle Stream (Pelletech) (liquid)	Liquid	< 95	U-D	< 47	E 12	< 12
Tertiary Pelletized Sludge (Pelletech)	Solid	*	E 220	E 6700	< 58	< 58

Notes: E = Estimated, ND = Not Detected, NA = Not Applicable, U-D = Result not reported because of sample interferences, Q-D = Result not reported because of result did not meet quality criteria, < = Denotes that the analyte was not detected; the associated parameter value is generally the reporting limit, * = Sample was not analyzed for this compound

Table 3-7. Plant B: Instantaneous Loads Results, YES Bioassay.

Sample Location	Eluent Fraction	Instantaneous Load (g/day, EE2 Eqs)			
		Apr-06	Jul-06	Oct-06	Jan-07
Primary Influent (L)	20 50 80	NA	13 T*	NA	3.4
Primary Influent (L) (Duplicate)	20 50 80	NA	16	NA	NA
Primary Influent (S)	20 50 80	NA	0.94	NA	0.034 T*
Primary Effluent (L)	20 50 80	NA	13	NA	0.26
Primary Effluent (L) (Duplicate)	20 50 80	NA	7.6 T*	NA	NA
Primary Effluent (S)	20 50 80	NA	0.12 T*	NA	0.26
Secondary Effluent (L)	20 50 80	NA	0.16	M	0.00019
Thickened Sludge (Combined Primary & Secondary)	20 50 80	0.082	NA	0.56	NA
Primary Unthickened Sludge	20 50 80	NA	0.084 T*	NA	NR/T
Secondary Unthickened Sludge	20 50 80	NA	0.047	NA	NR
Anaerobically Digested Sludge (Egg-shaped Digesters)	20 50 80	0.074 T*	0.15	0.47	0.063 T*

Notes: NR = no estrogenic response from sample, T = sample contained toxicity (no estrogenic response was observed), T* = sample contained toxicity (estrogenic response was also observed but was not quantified due to presence of toxicity), NA = not analyzed

Table 3-7. Plant B: Instantaneous Loads Results, YES Bioassay (continued).

Sample Location	Eluent Fraction	Instantaneous Load (g/day, EE2 Eqs)			
		Apr-06	Jul-06	Oct-06	Jan-07
Acid Phase Digested Sludge (L)	20 50 80	NA	NA	NA	0.00087
Acid Phase Digested Sludge (S)	20 50 80	NA	0.33	NA	0.31
Methane Phase Digested Sludge (L)	20 50 80	NA	NA	NA	NR
Methane Phase Digested Sludge (S)	20 50 80	NA	0.45	NA	T
Dewatered Sludge	20 50 80	0.084 T*	3.4	1.4	0.28
Centrate Recycle Stream (L)	20 50 80	0.065	0.16	0.11	0.012
Centrate Recycle Stream (L) (Duplicate)	20 50 80	0.22	NA	NA	NA
Centrate Recycle Stream (S)	20 50 80	0.012	0.014	0.0095	0.0070
Tertiary Pelletized Sludge	20 50 80	NA	NA	NA	0.0073
Composted Sludge	20 50 80	NA	NA	NA	0.077

Notes: NR = no estrogenic response from sample, T = sample contained toxicity (no estrogenic response was observed), T* = sample contained toxicity (estrogenic response was also observed but was not quantified due to presence of toxicity), NA = not analyzed

3.4.2 Chemical Analysis: Data Reduction Results and Discussion

3.4.2.1 Steroids

This section focuses on removal of steroids by activated sludge and various digestion processes.

One difficulty of comparing influent and effluent concentrations of TOrCs in wastewater treatment processes is that while such a comparison provides information on removal from the aqueous phase, no distinction can be made between chemical transformation of target compounds and physical removal by sorption to solids. The transformation of these compounds in the activated sludge process was assessed by comparing incoming load (primary effluent, or primary influent where a primary effluent sample is not available) with the total outgoing load (secondary effluent plus waste activated sludge).

None of the steroids were detected in the secondary unthickened sludge. However, the detection limit in this sample is severely constrained by the difficulty of extracting a large mass of solid material. The detection level varies inversely with the mass of sample extracted, so for extremely low mass samples (i.e., < 0.05 g dry weight), we were limited in our ability to assess removal. Therefore, we examined the activated sludge and thickening process as one unit, and the total outgoing load is now the load in secondary effluent plus the load in thickened sludge. There were not sufficient detections in the primary or secondary unthickened sludges to assess compound fate through the thickening process. This analysis shows that not only are estrogens removed from the aqueous stream relatively effectively, but that much of this removal is due to chemical transformation. Estriol exhibits greater than 99% removal, while E2 is transformed with 92% efficiency. The previously noted fact that E1 is a metabolite of E2 is likely the cause of reduced (66%) efficiency of E1 removal during this aerobic process. It should be noted that using this targeted chemical analysis; we cannot conclude that estrogenicity has been removed, as non-target metabolites could retain some activity. The sections discussing bioassay results provide insight as to the net effect on estrogenicity.

The non-estrogenic steroids also were transformed quite effectively. Often present in the influent at 10 to 100 times higher load than E2 or E1, all five androgens detected were transformed with 95% or greater efficiency. Three (dihydrotestosterone, testosterone, 11-ketotestosterone) were not detected at all in secondary effluent but all five had some residual signal in the solid stream. Progesterone, coprostanol, and cholesterol were removed from the liquid phase with greater than 98% efficiency, but there were significant residuals (15-20%) in the solid phases indicating less complete transformation.

Much like the aerobic digestion processes employed at Plant A, the aerobic activated sludge process has mixed effects on APEOs because of the potential interconversion of congeners with variable chain length. Total load of NP increases by approximately 40% while both NP1EO and NP2EO decrease during activated sludge treatment, with the majority of the load found in solid phase. Again, this is probably due to the degradation of longer chain NPEOs and formation of shorter chain NPEOs during aerobic treatment. Similarly, the load of OP increases while the load of OP1EO is decreasing. OP2EO was not detected in the influent so no evaluation can be made. Other EDCs of interest include bisphenol A, which decreases marginally, and β-sitosterol, which increases by a factor of 10.

As detailed in Section 2.1.2, Plant B incorporates a number of digestion processes. Since the thickened sludge stream is split between several digesters, the loads used for calculating process removal are flow weighted. Also, compounds that were well removed from the plant by upstream processes may have insufficient data for analysis of their behavior through digestion. Furthermore, many compounds are near enough to analytical detection levels that comparison of concentration changes may not be accurate. Based on our data, anaerobic digestion did a more effective job of removing target compounds than the two-stage acid and methane phase digestion. Conventional anaerobic digestion was only measured at one sampling time, so data are not sufficient to draw major conclusions here. Generally, the removal of steroid hormones from the plant was very good. Five androgens and three estrogens were detected in the plant influent (cis-androsterone, dihydrotestosterone, androstenedione, testosterone, 11-ketotestosterone, E1, E2, and E3). Of these, removal based on instantaneous loads was greater than 80% for all compounds except E1. As detailed previously, E1 is a known metabolite of E2 (Ternes et al., 1999) and likely E3. Average concentrations among numerous samplings are informative, but it is also instructive to examine trends during individual sampling periods. Although in many cases it was not possible to draw conclusions on seasonal variation, the behavior of E1 and the APEOs at plant B does offer some insight into seasonality. In the July 2006, E1 load to this plant was 57 g/day (primary effluent) and the total load going out (secondary effluent plus dewatered sludge) was 5.1 g/day, representing a decrease of over 90%. However, in January 2007, an incoming load of 14 g/day actually increased to 26 g/day in the combined liquid and solids streams (Figure 3-6). This is most likely because of effects of temperature on the kinetics of reaction 1. The transformation of E3 and E2 to final products is slower during the winter and more of the mass remains in the form of the intermediate (E1). Indeed, it is not possible to account for all of the E1 (26 g/day) leaving the plant solely from the incoming load of E2 (3.3 g/day) and E1 (14 g/day). Nevertheless, the compounds that exhibit near complete removal are minimally affected by seasonal differences in rate (Figure 3-7).

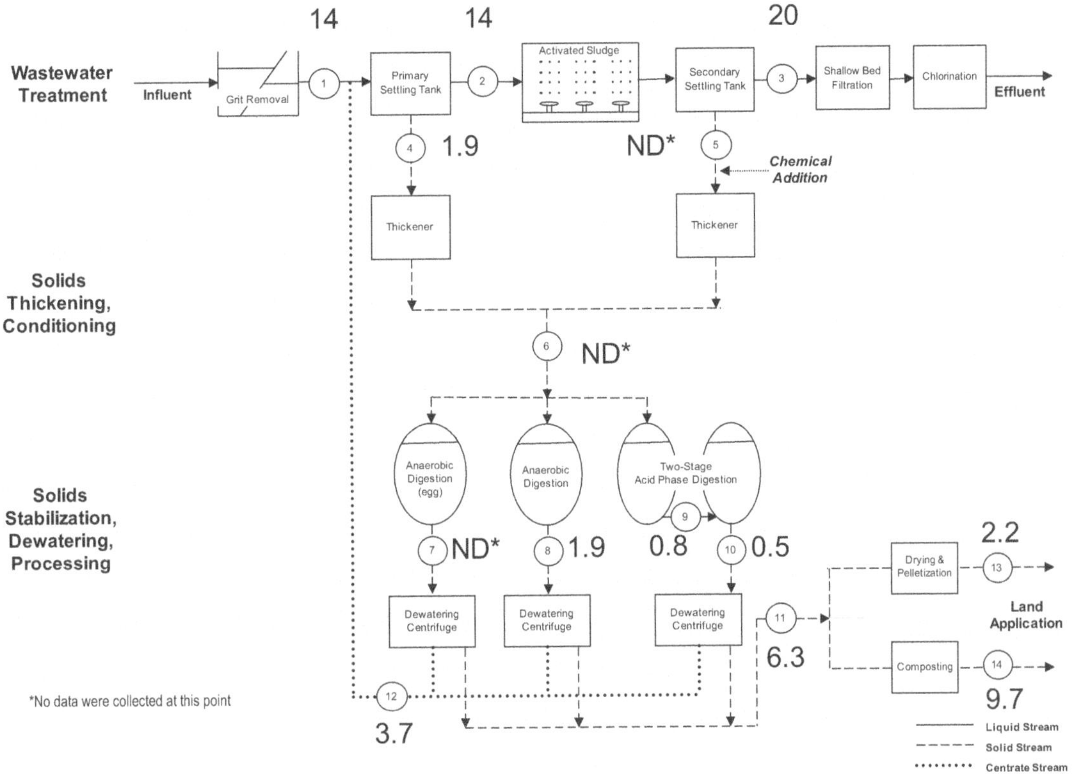

Figure 3-6. Estrone Flux (g/day) Through Plant B (January 2007).

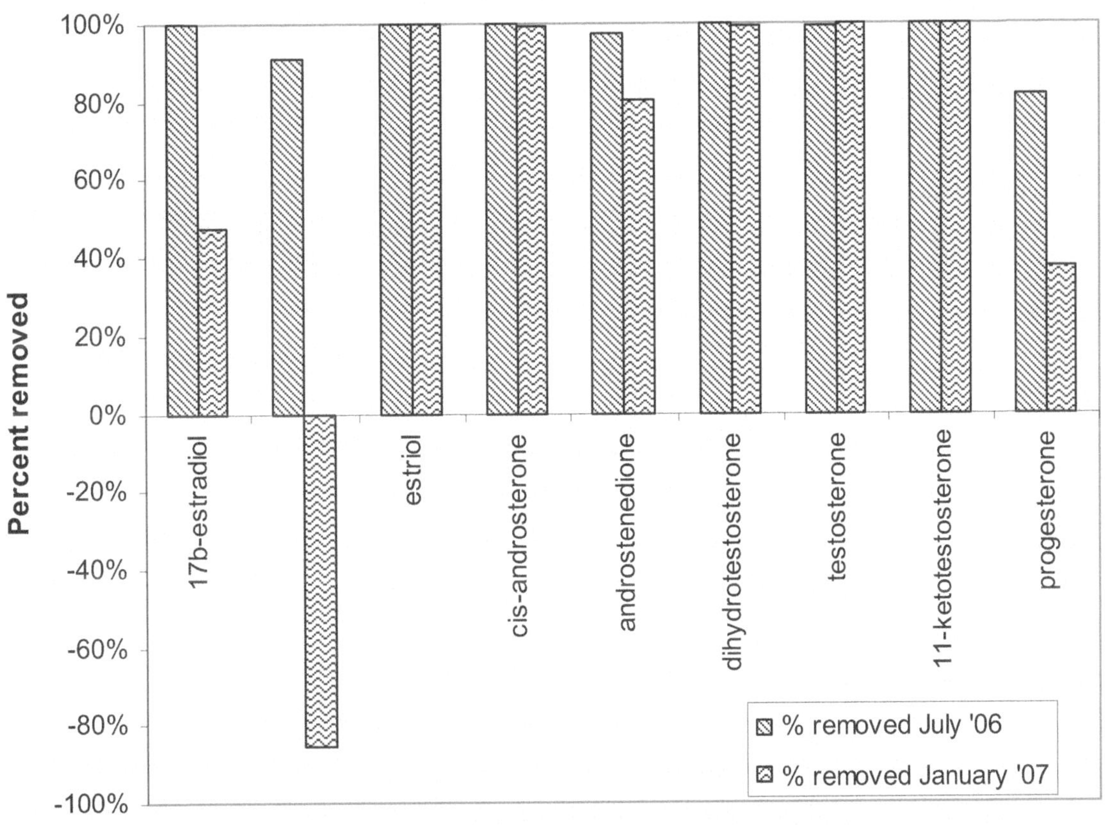

Figure 3-7. Plant B Hormone Removal, Seasonal Differences.

The incoming load of E3 (89 g/day) can more than account for this. This is the strongest evidence we have that E3 is being converted to E1 during the course of treatment. An alternative explanation is that incoming E1 and E2 are in the form of sulfate and glucoronide conjugates, and the mass increase is the result of deconjugation reactions, however available evidence would indicate that estrogens in WWTP influents are primarily in the deconjugated forms. A similar seasonal trend is also evident for the NPEs, with a sharp increase observed during January 2007 that is not apparent in the July 2006 (Figure 3-8). A similar kinetic explanation can be invoked here. Although longer chain APEOs were not measured during the course of this study, these surfactants are the source of the short chain APEOs that were measured. A slowing of the biological processes that degrade the long- and short-chain APEOs could impede carrying out of the process to the point of mineralization and result in higher concentrations of the intermediate degradation products (i.e. NP, NP1EO, NP2EO).

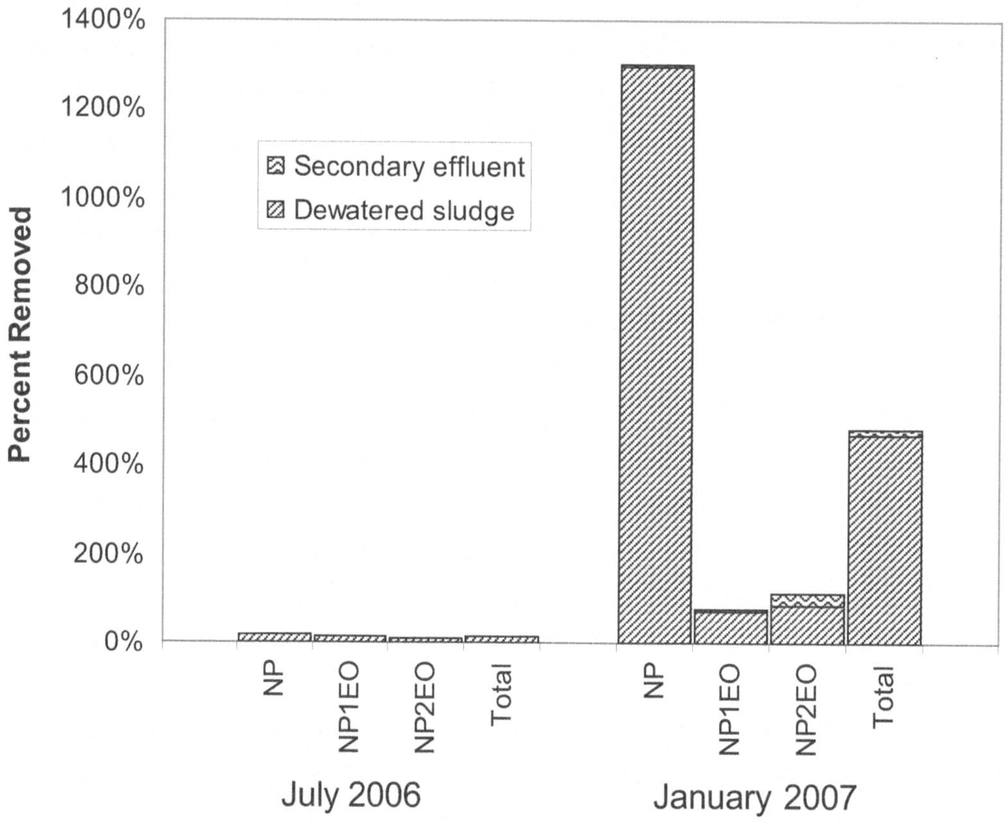

Figure 3-8. Plant B Differences in Alkylphenol Removal.

3.4.3 Biological Analysis: Data Reduction Results, Model of Concentration Addition, and Discussion

The sampling program and scope of the project did not permit a full investigation of estrogenic compound fate during all solids handling processes utilized at Plant B. The plant employs separate thickening processes that feed into three different anaerobic digestion processes: conventional and egg-shaped mesophilic anaerobic digesters, acid-phase digestion, and methane phase digestion.

An instantaneous loads analysis was performed around the egg-shaped digester. About 45% of the total thickened combined sludge mass flow was delivered to the egg-shaped digester. The other 55% entered the dual-phase acetogenesis/methanogenesis digester. Sample point 6 (thickened combined sludge) does not feed the four conventional anaerobic digesters (sample point number 7). They are supplied from two separate thickening operations, not shown.

To compute the instantaneous loads analysis around the egg shaped digester, the input from the thickened combined sludge mass flux was reduced to correspond to its mass flow fraction (45%).

Average mass flux rates of estrogenicity at Plant B, based on the Model of Concentration Addition, were determined for primary influent, primary effluent, secondary effluent, thickened combined sludge, and after mesophilic anaerobic digestion (egg-shaped digester only) (Figure 3-9). Similarly, average mass fluxes of estrogenic activity based on the YES bioassay results for those sample sampling points are shown in Figure 3-10. It is important to note that the sampling program at Plant B only permitted an instantaneous load analysis for the portion of sludge that was anaerobically digested in the egg-shaped digester. Thus, the mass fluxes in Figures 3-9 and 3-10 for the thickened combined sludge and the egg-shaped anaerobic digester represent only the fraction of solids going through the egg-shaped digestion process at Plant B.

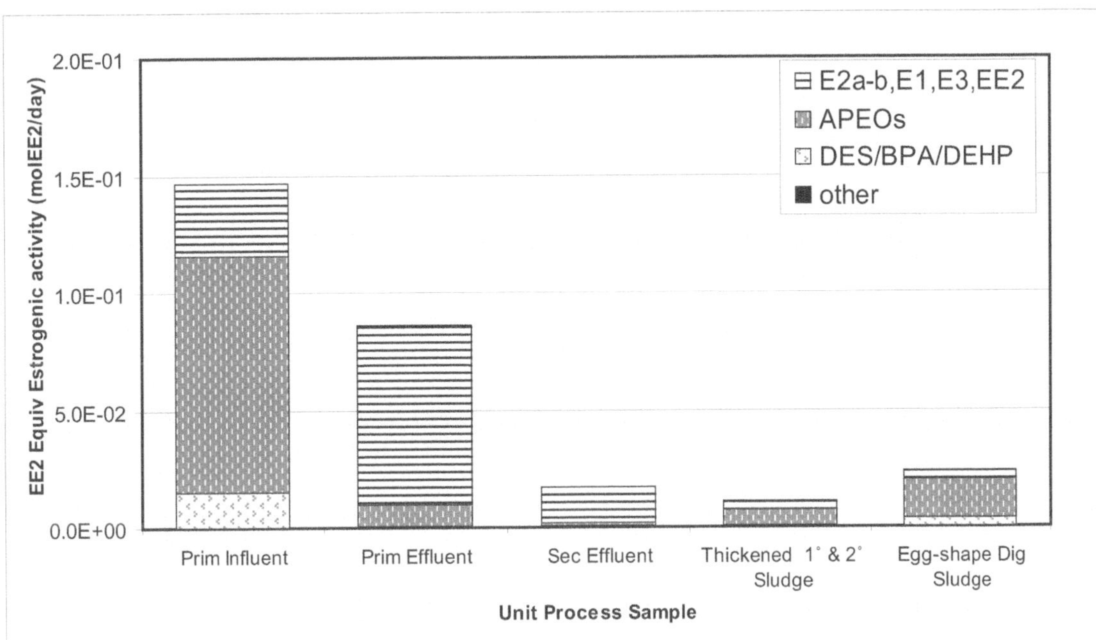

Figure 3-9. Average Daily Mass Fluxes (mol EE2-equivalents/day) of Estrogenicity at Plant B. (Based on the Model of Concentration Addition)

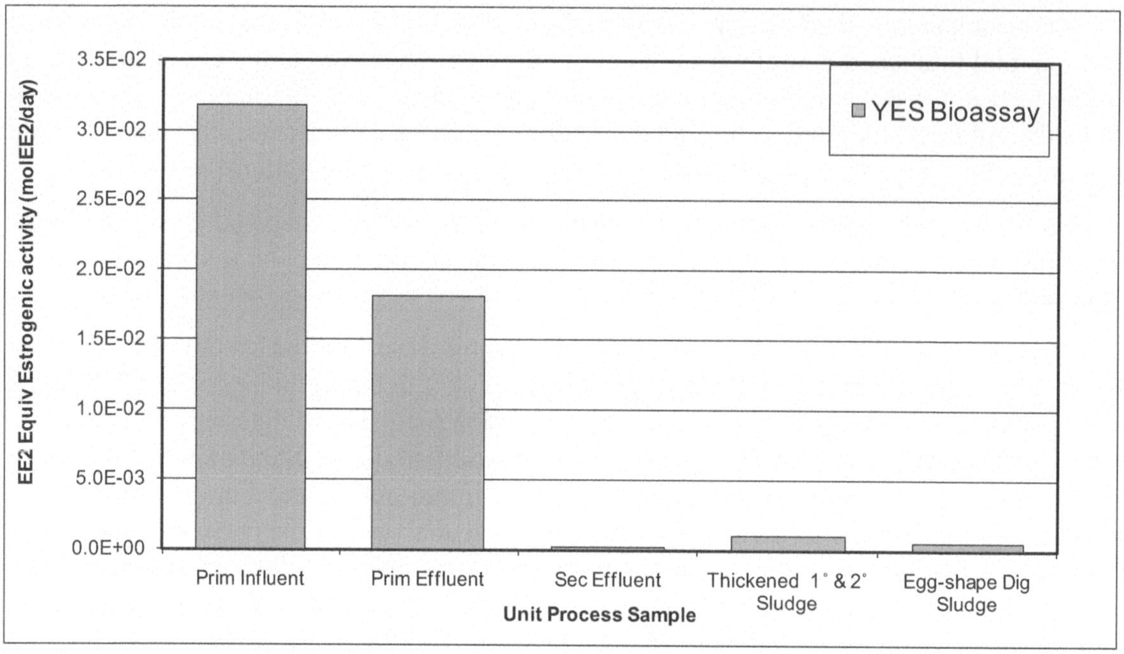

Figure 3-10. Average Daily Mass Fluxes (mol EE2-equivalents/day) of Estrogenic Activity at Plant B. (Based on YES Bioassay Measurements)

To evaluate the relationship of estrogenicity fate between liquid-stream and solid-stream samples, the contributing fluxes were reduced (normalized) to reflect the portion of flow associated with the loading of solids processed by the egg shaped digester. An assumption was made that uniform loading/distribution of target analytes among liquid and solid phases occurred during upstream treatment processes at Plant B. The adjusted mass fluxes for liquid-stream samples based on the Model of Concentration Addition and the YES bioassay results are shown in Figures 3-11 and 3-12, respectively. In Figure 3-11, estrogenic compounds are grouped as steroidal hormones, alkylphenols (APEOs), and other compounds, including DES and Bisphenol A. Total estrogenicity was substantially reduced during secondary treatment at Plant B. The amount of estrogenicity remaining in secondary effluent represented 13% of the total estrogenicity in primary influent. Steroidal hormones accounted for the majority of estrogenicity in primary and secondary effluents. There was a substantial net production of estrogenicity in the solids as a consequence of mesophilic anaerobic digestion. The increase is largely due to a greater contribution by APEOs. Nonylphenol in particular is important because it is created during the breakdown of aklylphenol polyethoxylates under anaerobic conditions and it has a much higher estrogenic potency than its longer chain, parent compounds. Overall, the YES bioassay measurements show similar trends as observed with the Model of Concentration Addition. There is a large decrease in estrogenicity during secondary treatment and the total estrogenicity of the solids increased after mesophilic anaerobic digestion. The mass flux of estrogenic activity increased from 0.486 to 0.638 mM EE2-equivalents/day, an increase of 31% during anaerobic digestion at Plant B. The chemical and bioassay measurements both reveal that there is a greater amount of estrogenicity discharged from this facility in the solids than in the secondary effluent.

The main hormone contributors to influent estrogenicity in Figure 3-11 were E2 (9% of total) and E1 (11.5% of total). After primary clarification, E2 comprised 38% of the total flux and E1 49%. Following secondary treatment, estrone contributed 70% of total remaining estrogenicity and E2 provided 20%. The average contribution to estrogenicity mass flux by nonylphenol in raw influent was 56%, decreasing to 9.5% in primary clarifier effluent and 1% in secondary effluent.

Bisphenol A was tested twice in raw influent and in January 2005 for primary effluent and secondary effluent. Bisphenol A and 4-tert octylphenol were about equal contributors to the influent estrogenicity (10% and 12% respectively), decreasing to less than 1% in primary effluent (Figure 3-11). Both BPA and 4-*tert* OP contributed 4% of the total estrogenicity mass flux after secondary treatment. Estriol, NP1EO, NP2EO, OP1EO, and OP2EO were < 1% for all liquid streams.

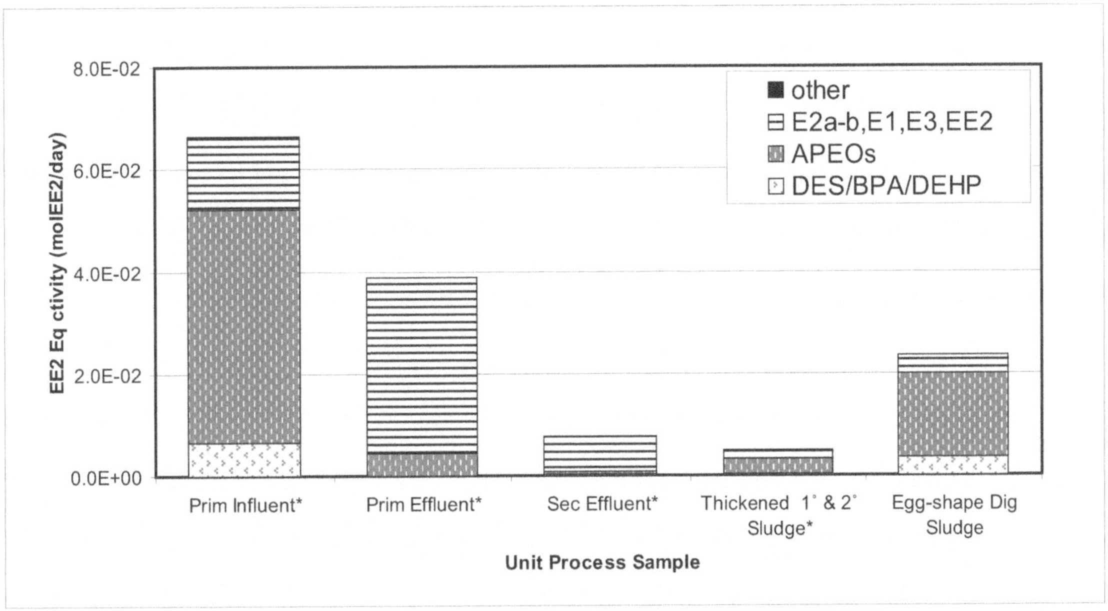

Figure 3-11. Average Daily Mass Fluxes (mol EE2-equivalents/day) of Estrogenicity at Plant B. (Based on the Model of Concentration Addition)

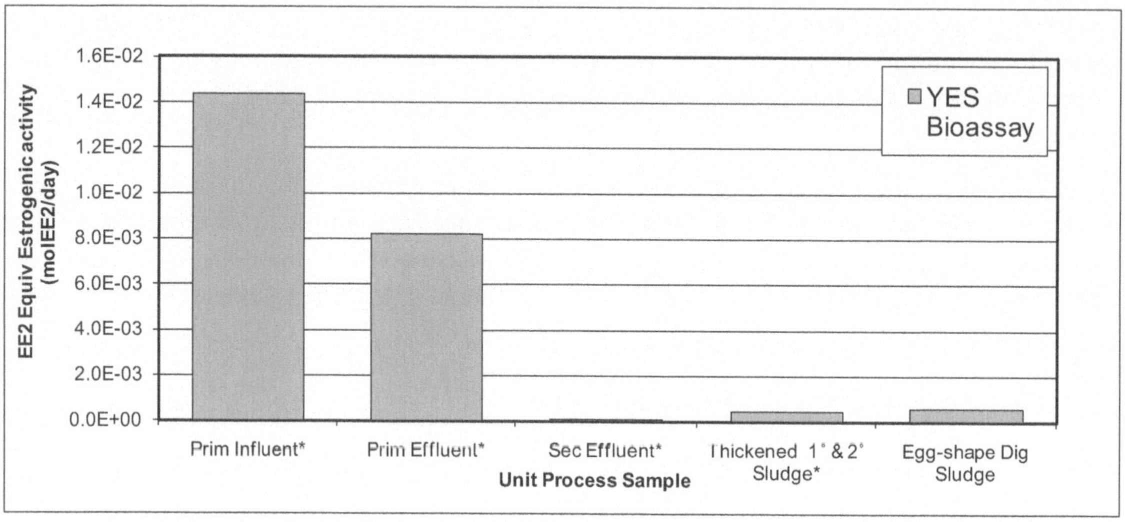

Figure 3-12. Average Daily Mass Fluxes (mol EE2-equivalents/day) of Estrogenic Activity at Plant B. (Based on YES Bioassay Measurements)

The magnitudes of estrogenic activity mass fluxes (based on the YES bioassay measurements) after unit treatment processes at Plant B are shown Figure 3-13 as percentages of the fluxes determined by the Model of Concentration Addition. That is, YES-based mass fluxes were normalized to corresponding mass fluxes based on the Model of Concentration Addition. YES-based mass fluxes varied from 2% (secondary effluent) up to 22% (raw influent) with the pre- and post-digested solids YES-based mass fluxes at 10% and 3%, respectively. Possible reasons for the lower response seen in the YES bioassay include presence of toxic and/or anti-estrogenic compounds, and competitive binding limitations. Samples from almost every sampling point at Plant B exhibited toxic effects during the YES bioassay tests. Toxicity was especially apparent for solid-phase samples from the egg-shaped anaerobic digester. Toxicity may be related to presence of sulfate or relatively high concentrations of DEHP as previously noted. Presence of anti-estrogens (antagonists) can block estrogenic response by preventing binding of sample estrogens to the human estrogen receptor during the YES bioassay (Conroy, 2005). Furthermore, there may be limits on the capacity of ligand-binding sites where estrogenic chemicals interact with the receptor and there may be a preference of certain estrogens over others (Terasaka, 2006).

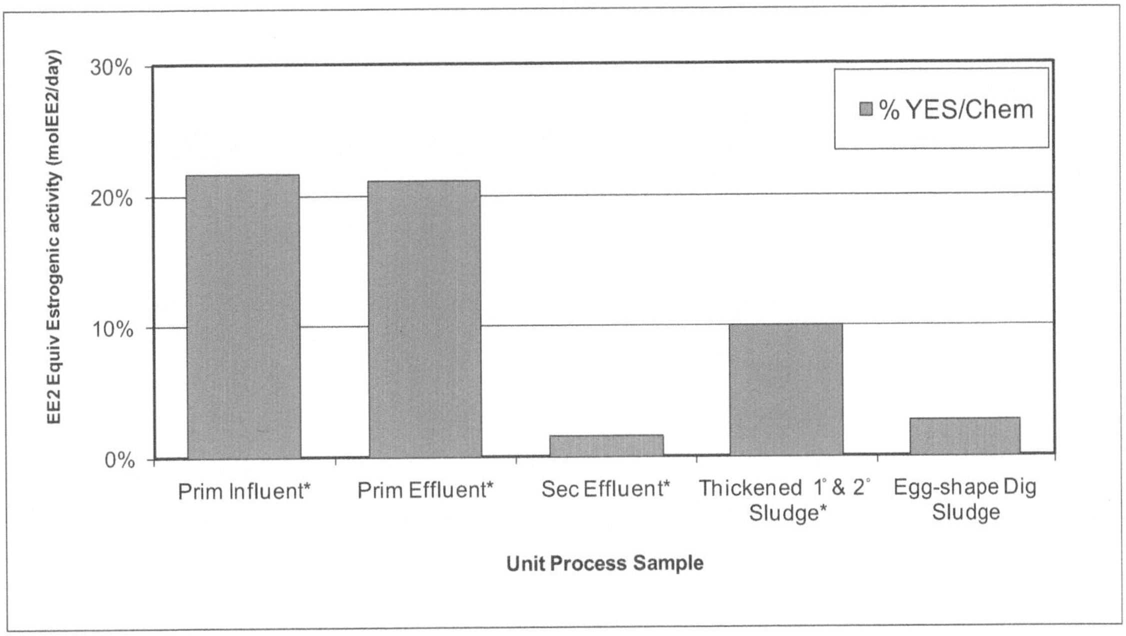

Figure 3-13. YES Bioassay Estrogenic Response as a % of Response Calculated Using the Model of Concentration Addition. (Calculated from Data shown in Figures 3-11 and 3-12)

The estrogenicity mass flux by APEOs increased from 3.125 to 16.574 mmol EE2-equivalents/day after mesophilic anaerobic digestion – an increase of 430%. In the pre-digested solids, nonylphenol contributed 58% of total mass flux, E1 (30%), 4-tert-OP (5%), BPA (3%), 17α-estradiol (2%), NP1EO (~1%) and DEHP (~1%). After mesophilic anaerobic digestion, NP contributed 65% of the mass flux; relative distributions of other alkylphenols were unchanged or decreased slightly: 4-t-OP (5%), all NP1-2EO and OP1-2EO compounds were <1%. DEHP was <1%, BPA increased (14%), and 17β-estradiol was detected in digested solids (5%) but was not detected prior to digestion.

The mass flux of estrogenicity contributed by estrogenic hormones increased from 1.537 to 3.422 mmol EE2-equivalents/day after mesophilic anaerobic digestion – an increase of 123%. The increase in hormones was unexpected. It was assumed that steroidal hormone activity would decrease due to conjugation with sulfate during anaerobic digestion. E1 contributed 9% of the remaining estrogenicity mass flux in the digested solids and 17α-estradiol provided <1%.

The relatively high nonylphenol and 4-*tert*-OP content in raw influent from Plant B coincided with relatively lesser amounts of shorter-chain NP and OP ethoxylates and suggests degradation of longer-chain ethoxylates occurred in the sewer before entering the treatment facility. The mass fluxes of shorter-chain ethoxylates (NP1EO, NP2EO, and OP1EO) decreased during mesophilic anaerobic digestion at Plant B, corresponding to increased fluxes of 4-*tert*-OP and NP. These results support the hypothesis of biodegradation of longer chain APnEOs (n=3-8) prior to entry into the treatment facility, leading to no corresponding rise in the AP(1-2)EO groups.

3.5 Plant C

3.5.1 Instantaneous Load: Hormones, Alkylphenolic Compounds, and Bioassays

Instantaneous loads calculations for all sample dates for steroid hormones and alkylphenolic compounds are provided in Table 3-8 and 3-9, respectively. Tables 3-10 and 3-11 show the instantaneous loads results for the YES bioassay at Plant C.

Table 3-8. Plant C: Instantaneous Loads Results (g/day), Hormones.

Sample Location	Matrix	diethylstilbestrol	cis-androsterone	epitestosterone	17-alpha-estradiol	dihydrotestosterone	androstene-3,17-dione	estrone	17-beta-estradiol	testosterone	equilin	11-ketotestosterone
December 2005												
Primary Waste Sludge	Solid	ND	E 410	ND	ND	ND	E 7.7	E 7.3	ND	ND	ND	E 2.7
Secondary Waste Sludge	Solid	ND	E 25	ND	ND	ND	E 10	E 47	ND	E 2.1	ND	E 10
Nit / Denit Waste Sludge	Solid	ND	E 0.10	ND	ND	ND	E 0.22	ND	ND	ND	ND	ND
Dewatered Sludge	Solid	3.1	77	ND	ND	7.3	22	7.7	ND	4.1	ND	ND
Centrate Recycle (Liquid)	Liquid	ND	ND	ND	1.8	ND	ND	ND	0.77	0.17	ND	ND
Centrate Recycle (Solid)	Solid	ND	1.2	ND	ND	ND	ND	0.44	ND	ND	ND	ND
Lime Stabilized Sludge	Solid	ND	41	ND	ND	7.7	5.8	3.0	3.0	ND	ND	1.9
Secondary Waste Sludge (Duplicate)	Solid	ND	25	ND	ND	ND	ND	ND	ND	ND	ND	ND
July 2006												
Dewatered Sludge	Solid	ND	39	ND	ND	ND	15	1.8	2.0	ND	ND	1.0
Centrate Recycle (Liquid)	Liquid	D-R	D-R	D-R	D-R	D-R	D-R	D-R	D-R	D-R	D-R	D-R
Centrate Recycle (Solid)	Solid	ND	E 200	ND	ND	3.0	42	28	ND	13	ND	ND
Lime Stabilized Sludge	Solid	ND	38	ND	ND	2.6	4.3	1.1	0.95	ND	ND	0.87
Lime Stabilized Sludge (Duplicate)	Solid	ND	53	ND	ND	2.9	4.3	0.82	1.6	ND	ND	0.68
Lime Stabilized Sludge (Triplicate)	Solid	16	49	ND	ND	7.0	5.8	1.7	0.96	2.4	ND	ND

Notes: E = Estimated, ND = Not Detected, NA = Not Applicable, U-D = Result not reported because of sample interferences, D-R = Value is not reported because sample or analyte was ruined. Often due to matrix issues., Q-D = Result not reported because of result did not meet quality criteria, < = Denotes that the analyte was not detected; the associated parameter value is generally the reporting limit, * = Sample was not analyzed for this compound

Table 3-8. Plant C: Instantaneous Loads Results (g/day), Hormones (continued).

Sample Location	Matrix	19-norethindrone	mestranol	equilenin	17-alpha-EE2	estriol		progesterone		coprostanol		cholesterol		Total	Without coprostanol and cholestorol	Without cholesterol
December 2005																
Primary Waste Sludge	Solid	ND	ND	ND	ND	E	4.1	ND		E	240	E	260	930	430	670
Secondary Waste Sludge	Solid	ND	ND	ND	ND	E	1.8	ND		E	60	E	60	220	97	160
Nit / Denit Waste Sludge	Solid	ND	ND	ND	ND	ND		ND		E	0.26	E	0.26	0.84	0.32	0.58
Dewatered Sludge	Solid	ND	ND	ND	ND		1.1		15	E	300	E	470	910	140	440
Centrate Recycle (Liquid)	Liquid	ND	ND	ND	ND		8.2	ND			59		36	110	11	70
Centrate Recycle (Solid)	Solid	ND	ND	ND	ND	ND		ND			3,200		3,300,000	3,300,000	1.6	3,200
Lime Stabilized Sludge	Solid	ND	ND	ND	ND	ND		ND		E	350	E	380	800	62	420
Secondary Waste Sludge (Duplicate)	Solid	ND	ND	ND	ND	D-R		ND		E	240	E	240	500	25	260
July 2006																
Dewatered Sludge	Solid	ND	ND	ND	ND	ND			4.8	E	290	E	390	750	63	360
Centrate Recycle (Liquid)	Liquid	D-R	D-R	D-R	D-R	D-R		D-R		D-R		D-R		D-R		
Centrate Recycle (Solid)	Solid	ND	ND	ND	ND	ND			19	E	180	E	36	520	300	480
Lime Stabilized Sludge	Solid	ND	ND	ND	ND	ND			4.2	E	360	E	380	780	52	410
Lime Stabilized Sludge (Duplicate)	Solid	ND	ND	ND	ND	ND			4.7	E	190	E	240	500	68	260
Lime Stabilized Sludge (Triplicate)	Solid	ND	ND	ND	ND		1.7		5.4	E	4,900	E	530	5,500	90	5,000

Notes: E = Estimated, ND = Not Detected, NA = Not Applicable, U-D = Result not reported because of sample interferences, D-R = Value is not reported because sample or analyte was ruined. Often due to matrix issues., Q-D = Result not reported because of result did not meet quality criteria, < = Denotes that the analyte was not detected; the associated parameter value is generally the reporting limit, * = Sample was not analyzed for this compound

Table 3-9. Plant C: Instantaneous Loads Results (g/day), Alkylphenolic Compounds.

Sample Location	Matrix		4-Nonylphenol		4-Nonylphenol monoethoxylates		4-Nonylphenol diethoxylates		4-ter-octylphenol		4-Octylphenol monoethoxylates		4-Octylphenol diethoxylates		Bisphenol A		beta-Sitosterol		1,4-Dichlorobenzene		4-Cumylphenol
December 2005																					
Primary Waste Sludge	Solid	E	180,000		92,000	E	48,000	<	3.8	E	9,100	<	7.6	E	8,200		41,000	<	3.8	<	3.8
Secondary Waste Sludge	Solid	E	250,000		71,000	E	29,000		2,600	<	6.5	<	6.5	<	6.5		38,000	<	3.3	<	3.3
Nit / Denit Waste Sludge	Solid	E	670		760	E	360	<	0.26	<	0.52	<	0.52	<	0.52		2,100	<	0.26	<	0.26
Dewatered Sludge	Solid	E	150,000		25,000	<	22,000	<	6.2	<	12	<	12	<	12		11,000	<	6.2	<	6.2
Centrate Recycle	Liquid	E	4.5	*		<	31	<	6.2	<	6.2	<	6.2	U-D		<	12	<	3.1	<	6.2
Lime Stabilized Sludge	Solid	E	170,000		44,000	E	46,000	<	6.8	<	14	E	440	<	14		14,000	<	6.8	<	6.8
Dewatered sludge (duplicate)	Solid	E	150,000		32,000	E	35,000	<	6.2	<	12	<	12	<	12		11,000	<	6.2	<	6.2
July 2006																					
Dewatered Sludge	Solid		2,200,000		4,600,000		710,000		42,000		74,000	<	16,000	U-D			3,600,000		4,600	<	16,000
Lime Stabilized Sludge	Solid		150,000		380,000		100,000		3,100		7,300	<	1,900	U-D			380,000		250	<	1,900
Centrate Recycle	Liquid	E	7.3	*		<	56	E	0.21	E	0.35	<	11	E	1.8	<	23	E	0.22	<	11

Notes: E = Estimated, ND = Not Detected, NA = Not Applicable, , U-D = Result not reported because of sample interferences, Q-D = Result not reported because of result did not meet quality criteria, < = Denotes that the analyte was not detected; the associated parameter value is generally the reporting limit, * = Sample was not analyzed for this compound

Table 3-10. Plant C: Instantaneous Loads Results, YES Bioassay (December 2005).

Sample Location	Eluent Fraction	Instantaneous Load (g/day, EE2 Eqs)
Primary Waste Sludge	20	T
	40	
	60	
	80	
Secondary Waste Sludge	20	T
	40	
	60	
	80	
Nit/Denit Waste Sludge	20	0.37
	40	
	60	
	80	
Dewatered Sludge	20	2.5
	40	
	60	
	80	
Centrate Recycle (Liquid)	20	0.0009
	80	
	100	
Centrate Recycle (Solid)	20	0.55
	40	
	60	
	80	
Lime Stabilized Sludge	20	9.7
	40	
	60	
	80	

Notes: NR = no estrogenic response from sample, T = sample contained toxicity (no estrogenic response was observed), T* = sample contained toxicity (estrogenic response was also observed but was not quantified due to presence of toxicity), NA = not analyzed

Table 3-11. Plant C: Instantaneous Loads Results, YES Bioassay (July 2006).

Sample Location	Eluent Fraction	Instantaneous Load (g/day, EE2 Eqs)
Dewatered Sludge	20	6.2
	50	
	80	
Centrate Recycle (Liquid)	20	0.028
	50	
	80	
Centrate Recycle (Solid)	20	1.0
	50	
	80	
Lime Stabilized Sludge	20	26
	50	
	80	

Notes: NR = no estrogenic response from sample, T = sample contained toxicity (no estrogenic response was observed), T* = sample contained toxicity (estrogenic response was also observed but was not quantified due to presence of toxicity), NA = not analyzed

3.5.2 Chemical Analysis: Data Reduction Results and Discussion

3.5.2.1 Steroids

Plant C was selected for this study largely because it incorporates a lime stabilization process after sludge dewatering. As such, samples were not collected from the liquid streams and only at sites meant to understand the conversion of compounds through the lime stabilization process. One consequence of this decision is that many compounds that are typically well removed by activated sludge treatment were not present in any of the samples. Another consequence of lime stabilization is that due to the extremely high pH, which exceeded the pKa of many phenolic compounds, they were present in the lime stabilized sludge in both protonated and deprotonated form. This could have implications on both activity and removal, although sample extracts are buffered in phosphate, so it should not impact the chemical analysis.

Of the steroids, only estrone, androstenedione, and cis-androsterone were present with sufficient frequency for a complete analysis. The lime stabilization process actually appears to be more effective during the winter sampling period (December 2005) than in the summer (July 2006), although one winter and one summer sampling event may not be sufficient density to draw firm conclusions. Cis-androsterone is not substantially removed in the summer but load decreases by nearly 50% in the winter Estrone load decreases by 60% in the winter, versus only 40% in the summer, and androstenedione removals are seasonally comparable (70% vs. 73%) (Table 3-12). It is interesting to note that the measured load of all three of these compounds is substantially larger in the centrate recycle stream in the summer, but absent from the stream in the winter. This could be indicative of a greater tendency towards sorption during the colder periods.

Table 3-12. Plant C Steroid Removal, Seasonal Differences.

Test Compound	% Removal, Lime Stabilization (December 2005)	% Removal, Lime Stabilization (July 2006)
diethylstilbestrol	100%	ND
cis-androsterone	47%	3.9%
dihydrotestosterone	-5.1%	ND
androstene-3,17-dione	73%	71%
estrone	61%	39%
17-beta-estradiol	ND	53%
testosterone	100%	ND
11-ketotestosterone	ND	13%
estriol	100%	ND
progesterone	100%	13%
coprostanol	-19%	-22%
cholesterol	19%	3.9%

Note: ND = Not Detected

3.5.2.2 Non-Steroidal Estrogenic Compounds

The lime stabilization process is highly effective at removing non-steroidal estrogenic compounds during both summer and winter. NP plus NP2EO is nearly 3,000,000 g/d in July and is reduced by over 80% during the process. Although APEOs are 1,000 to 100,000 times less potent than the steroids (i.e., 30 to 3,000 g EE2-equivalents/d in dewatered sludge), only estrone (7.7 g/day, 2.5 g EE2-equivalents/d) was detected in the dewatered sludge which feeds the lime stabilization process. Therefore, it appears that in this case APEOs account for most of the estrogenicity present and the lime stabilization process has caused significant removal of total estrogenicity prior to solids disposal.

3.5.3 Biological Analysis: Data Reduction Results, Model of Concentration Addition, and Discussion

The final solids stabilization process at Plant C consists of lime addition to thickened combined primary and secondary sludge. The secondary treatment process includes nutrient removal (nitrification/denitrification), which has been shown previously to greatly reduce many estrogenic compounds.

Samples were collected twice: December 2005 and July 2006. The average (n = 2) mass flux of estrogenicity of the thickened combined solids at Plant C decreased by 93% after lime stabilization, based on the Model of Concentration Addition (Figure 3-14). The estrogenicity mass flux in the thickened combined sludge was dominated by APEOs. Nonylphenol contributed 87% of total estrogenicity, 4-tert-OP provided 9%, and NP1EO, NP2EO and OP1EO together provided 2%. 4-n-octylphenol (4-n-OP) and 4-octylphenol diethoxylate (OP2EO) were not detected. The steroidal hormones contributed little estrogenicity to the dewatered combined sludge with 17β-estradiol and estrone accounting for approximately 1% of the total estrogenicity mass flux. 17-α estradiol and 17-α ethynyl estradiol were not detected in either sampling period. Likewise, DES contributed less than 1%.

The estrogenicity mass flux remaining after lime stabilization was largely attributed to APEOs with nonylphenol contributing 65% of the estrogenicity (Figure 3-14). Total mass flux of APEOs decreased from 812 to 54.6 M EE2-equivalents/day. The relative distribution of most APEOs was unchanged by lime addition. The mass flux of estrogenic hormones (E2-α, E2-β, E1, EE2, and E3) increased by 6.5% after lime stabilization (Figure 3-15). E1 and E2 combined to 13% of total remaining estrogenicity. The mass flux of DES during lime stabilization varied. A reduction occurred in the December 2005 and an increase in July 2006. On average, there was a net flux increase of 72% for DES (see Section 3.2.1).

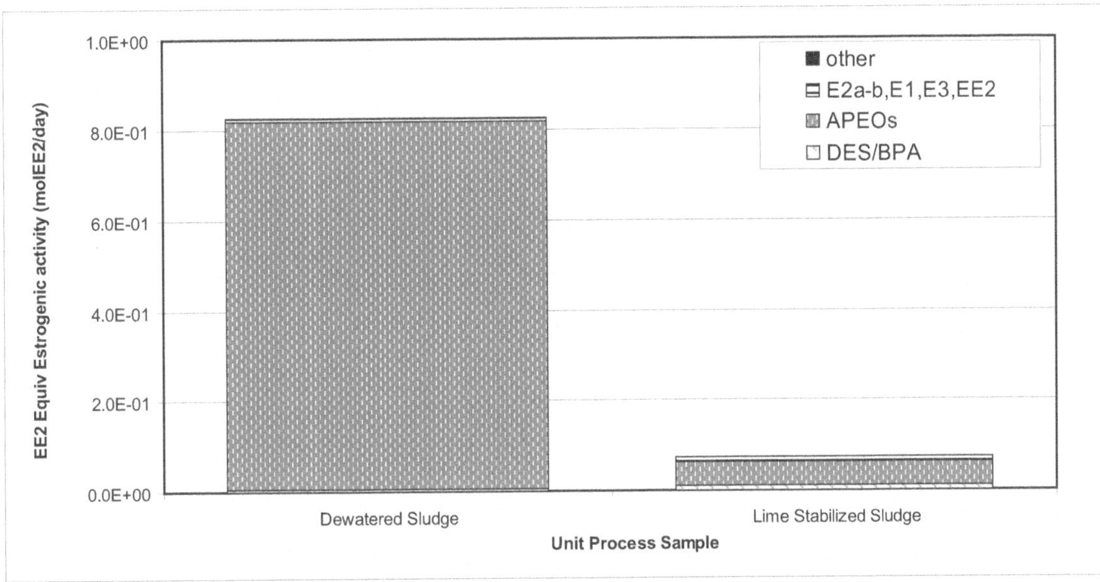

Figure 3-14. Average Daily Mass Fluxes (mol EE2-equivalents/day) of Estrogenicity Before and After Lime Stabilization at Plant C. (Based on the Model of Concentration Addition)

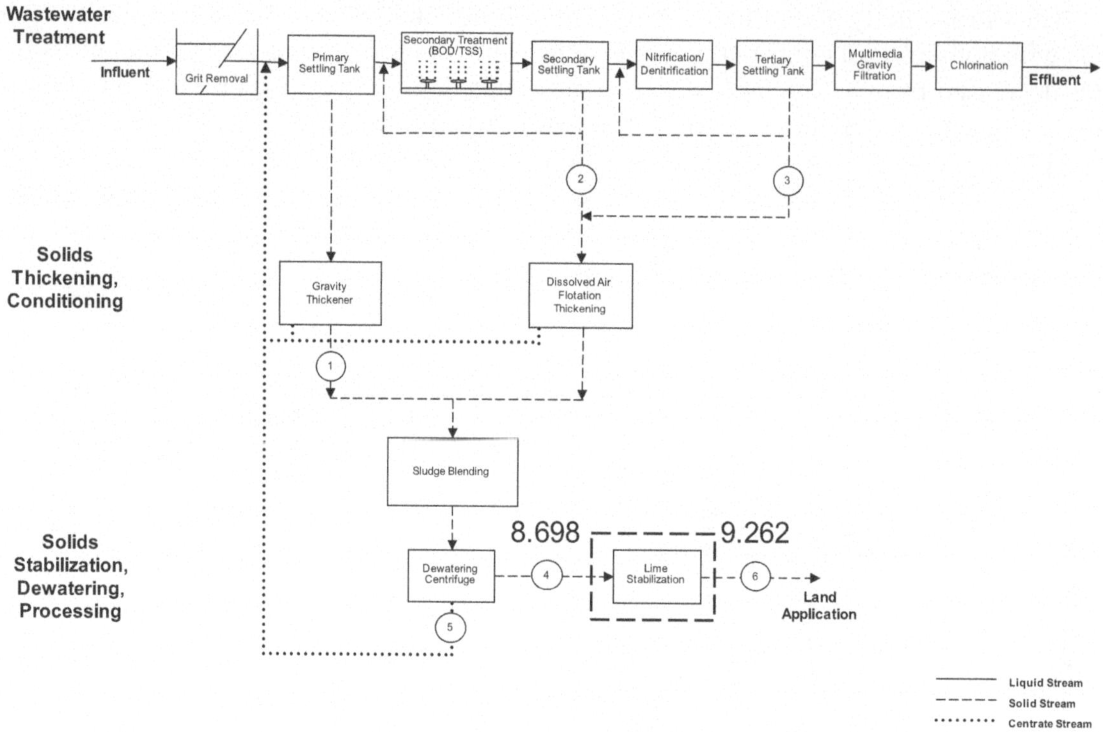

Figure 3-15. Daily Estrogenicity Mass Flux (mmol EE2-equivalents/day) Due to Estrogenic Hormones (E2-α, E2-β, E1, EE2 and E3) Before and After Lime Stabilization at Plant C. (Based on the Model of Concentration Addition)

Based on the bioassay, the average (n = 2) mass flux of estrogenic activity before and after lime addition increased by 312% (Figure 3-16). Results from the December 2005 and July 2006 measurements were similar, with increases of 286% and 322%, respectively. The magnitude of estrogenic activity mass fluxes (based on the YES bioassay measurements) for the pre- and post-lime stabilized solids represented 2% and 83%, respectively, of the mass fluxes based on the Model of Concentration Addition.

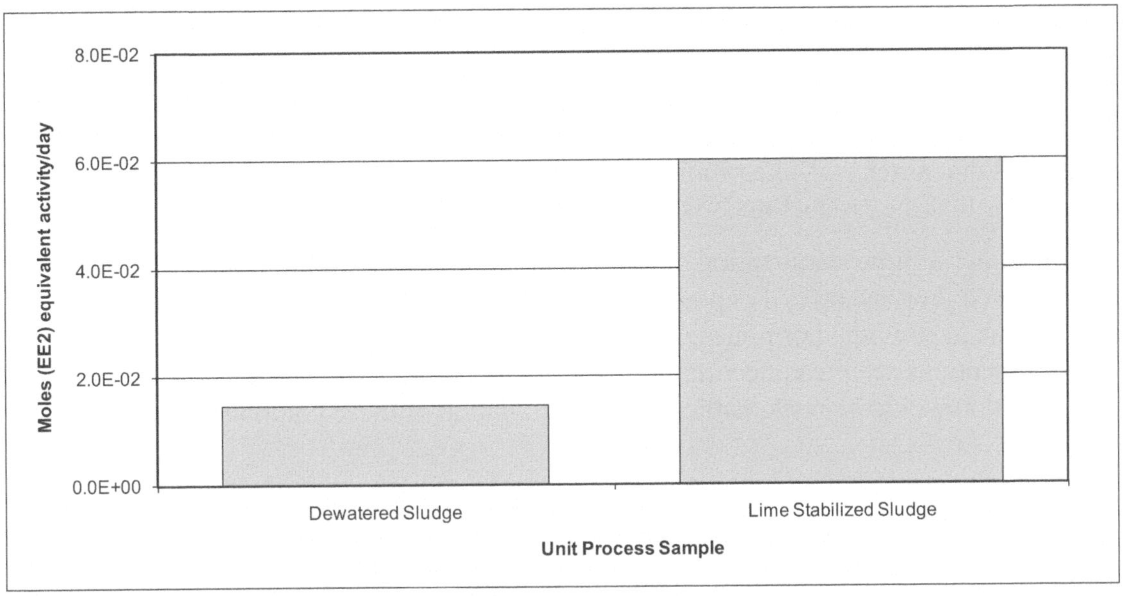

Figure 3-16. Daily Average Mass Fluxes of Estrogenic Activity Before and After Lime Stabilization at Plant C. (Based on YES Bioassay Measurements)

The most estrogenic steroidal hormones reported in Table 3-8 are 17-alpha-estradiol (17-alpha-E), E1, E2, E3, and EE2. The concentrations of these compounds and the bioassay-derived total estrogenicity for both the December 2005 and July 2006 sample collection dates are provided in Tables 3-10 and 3-11. Within the uncertainties associated with measurements of solids, chemical analysis of hormones, and bioassay analysis of total estrogenicity, the sum of the concentrations of these five analytes is reasonably comparable to the YES bioassay results.

Table 3-9 provides the results for alkylphenolic compounds for the July 2006 sample collection date. The bulk of estrogenicity in aqueous treatment streams results from relatively few compounds; the major contributions come from the steroids E2, EE2, E1, and to a lesser extent, E3. Alkylphenols, alkylphenol ethoxylates, bisphenol A, and other non-steroidal estrogenic compounds are typically present in treated effluents at ug/L levels (compared to ng/L for the hormones). However, their relative activity is such that outside of a few well-documented special cases their contribution to total estrogenicity of effluents is relatively small. However, alkylphenols are somewhat more hydrophobic than the steroids, so the relative importance of this chemical class in the solid phase should be somewhat higher than in the liquid phase. Thus, it is expected that the total estrogenicity of sludges may result more from the alkylphenolic compounds than from the steroid hormones. Table 3-9 shows that, with the exception of bisphenol A, concentrations of each compound were reduced between 88 to 93% following lime stabilization. Further, in the July 2006 samples, the centrate stream had the highest concentrations of steroidal compounds while the nonylphenolic compounds had the lowest concentrations found in the centrate stream.

In contrast to the decrease of most monitored estrogenic compounds, there was an approximate four-fold increase in estrogenic activity following lime stabilization, as measured by the YES bioassay.

Lime stabilization is widely used to stabilize sludges. In this process, lime is added to untreated sludge in sufficient quantity to raise the pH to 12 or higher. The high pH creates an environment that halts or substantially retards the microbial reactions that can otherwise lead to odor production and vector attraction (Metcalf and Eddy, 2003). The sludge will not putrefy, create odors, or pose a health hazard so long as the pH is maintained at this level. The process can also inactivate virus, bacteria, and other microorganisms present. An advantage of lime stabilization is that a rich soil-like product results with substantially reduced pathogens. A disadvantage is that the product mass is increased by the addition of the lime material.

The effect of lime stabilization on compound concentrations and estrogenicity needs to be further evaluated. For instance, it is possible that a dilution effect due to lime addition accounts for the decrease in steroidal hormone concentrations. Dilution does not likely account for the significant reductions seen in concentrations of the alkylphenolic compounds but it is not likely that lime addition would be responsible for the degradation of these compounds to the extent observed.

Although raising pH to 12 has significant benefits to sludge quality as outlined above, it also drastically changes the chemistry of many of the microconstituents being analyzed and the nature of the solids phase itself. Estrone has a pK_a of approximately 10.4 and the other hormones and alkylphenols fall in a similar range. Many of the pharmaceuticals have pK_a's that are much lower and may be less affected by the pH increase during lime stabilization. Relative to samples at lower pH, a greater fraction of the compounds in lime-stabilized sludge will be in a deprotonated (anionic) form, and the solid phase will have a strongly negative surface charge. At a pH of 12, 97.5% of estrone should be deprotonated (compared to < 0.1% at neutral pH). We have indirect evidence that this phenomenon is occurring derived from the fractionation approach applied to the bioassay samples. In July 2006, greater than 90% of the estrogenicity in lime-stabilized sludge eluted in the 20% methanol fraction, compared to less than 3% in the dewatered sludge which would have a lower pH. This suggests that compounds responsible for the estrogenicity were more polar in nature for the lime-stabilized sludge.

There are at least two potential explanations for the observation that estrogenicity increases after lime stabilization while concentrations of individual estrogens do not. Both of these have their basis in the decrease in the importance of hydrophobic partitioning expected when the surface of the solid phase and individual microconstituents simultaneously become increasingly polar. First, it is possible that a strongly sorbed fraction of the estrogenic compounds becomes more available to extraction and cleanup. This is unlikely to be the cause of the observed difference because analyte recoveries were generally good on samples without lime addition, and because USGS techniques require addition of pH 7 buffer to sample extracts prior to the cleanup steps in the methods, which are designed to remove polar interferences from the matrix. However, methods were validated using samples at more typical environmental pH so we will verify that the added buffering capacity is sufficient to lower the pH to near neutral prior to extraction. Another possibility is that the estrogenicity derives from non-target analytes or matrix components that become more available to extraction via a similar mechanism of decreased partitioning.

It is possible the improved agreement between the two methods after lime stabilization is due to a suggested transformation whereby the addition of a strong base such as lime (calcium hydroxide) may transform some compounds into more estrogenic forms that were not included in the analyzed suite of compounds. For example, aromatic epoxides can be non-enzymatically hydroxylated into ketones and phenols (Dowers, 2004). Addition of an unhindered phenolic OH group in a *para* position to compounds with molecular weights between 140-250 Daltons is known to increase estrogenic response in the YES bioassay (Miller, 2001). Phenolic hydroxylation of benzophenone and other proestrogens by a common enzymatic catalyst in the cyctochrome P450 system can convert these compounds into estrogenic forms (Kawamura, 2003; Kitamura, 2008).

At high pH, nonylphenol and 4-tert-octylphenol can be transformed to longer chained APEOs in the presence of ethylene oxide (Zoller, 2009). Addition of lime may have created longer ethoxylate-chained alkylphenols (that were not target analytes in this project) from precursor NP and OP compounds with short-chains or absent EO groups. If this transformation did occur, it could partially account for the substantial decreases in nonylphenol and octylphenol mass fluxes at Plant C after lime stabilization. Finally, hydroxylation of proestrogens (inactive) could have transformed inactive compounds into estrogenically active compounds not measured in this project.

3.6 Plant D

3.6.1 Instantaneous Load: Hormones, Alkylphenolic Compounds, and Bioassays

Instantaneous loads calculations for all sample dates for steroid hormones, and alkylphenolic compounds are provided in Tables 3-13 and 3-14, respectively. Table 3-15 shows the instantaneous loads results for the YES bioassay at Plant D.

Table 3-13. Plant D: Instantaneous Loads Results (g/day), Hormones.

Sample Location	Matrix	diethylstilbestrol	cis-androsterone	epitestosterone	17-alpha-estradiol	stanolone	androstene-3,17-dione	estrone	17-beta-estradiol	testosterone	equilin
March 2006											
Primary Sludge (Unthickened)	Solid	ND	110	ND	ND	14	23	9.1	6.0	8.4	ND
Thickened Wasted Activated Sludge (TWAS)	Solid	ND	E 3.6	ND	E 0.52	ND	E 2.0	E 6.2	ND	E 0.33	ND
Dewatering Centrate	Solid	ND	0.21	ND	ND	ND	0.31	ND	ND	ND	ND
Digested Sludge	Solid	ND	E 2,800	ND	E 3.3	E 33	E 28	E 130	ND	E 6.6	ND
Digested Sludge (Duplicate)	Solid	ND	E 2,300	ND	E 2.6	E 31	E 28	E 110	ND	E 6.6	ND
June 2006											
Primary Influent	Liquid	< 1.0	E 6,900	78	21	160	E 720	94	53	150	< 5.0
Primary Effluent	Liquid	< 1.0	E 11,000	86	< 1.0	180	490	77	33	120	< 5.2
Secondary Effluent	Liquid	< 1.0	9.1	< 5.0	0.74	< 1.0	44	6.2	1.5	1.5	< 5.0
Primary Sludge (Unthickened)	Solid	9.5	220	ND	ND	16	7.1	15	6.4	ND	ND
Waste Activated Sludge (Unthickened WAS)	Solid	ND	17	ND	4.0	ND	8.9	20	ND	ND	ND
Thickened Wasted Activated Sludge (TWAS)	Solid	ND	58	ND	6.4	ND	7.4	11	ND	ND	ND
Thickened Wasted Activated Sludge (TWAS) (Duplicate)	Solid	ND	46	ND	6.0	3.0	6.6	12	ND	ND	ND
Thickened Wasted Activated Sludge (TWAS) (Triplicate 1)	Solid	ND	6.6	ND	1.1	ND	4.0	4.1	2.9	ND	ND
Thickened Wasted Activated Sludge (TWAS) (Triplicate 2)	Solid	ND	21	ND	1.9	ND	4.8	13	ND	ND	ND
Dewatering Centrate	Liquid	0.021	E 31	< 0.20	0.29	0.31	2.9	E 4.6	< 0.039	0.095	< 0.20
Dewatering Centrate	Solid	ND	2.9	0.73	ND	1.3	17	2.1	0.11	2.6	ND
Digested Sludge	Solid	0.23	100	ND	0.63	1.2	3.3	12	0.74	ND	ND
TWAS Centrate	Liquid	< 0.13	0.77	< 0.67	2.2	< 0.13	0.51	1.6	0.48	0.25	< 0.67
TWAS Centrate	Solid	ND	28	ND	0.88	ND	0.97	3.0	0.91	ND	ND
TWAS Centrate (Duplicate)	Solid	ND	42	ND	ND	ND	6.7	ND	1.7	ND	ND

Notes: E = Estimated, ND = Not Detected, NA = Not Applicable, U-D = Result not reported because of sample interferences, D-R = Value is not reported because sample or analyte was ruined. Often due to matrix issues., Q-D = Result not reported because of result did not meet quality criteria, < = Denotes that the analyte was not detected; the associated parameter value is generally the reporting limit, * = Sample was not analyzed for this compound

Table 3-13. Plant D: Instantaneous Loads Results (g/day), Hormones (continued).

Sample Location	Matrix	11-ketotestosterone	19-norethindrone	mestranol	equilenin	17-alpha-EE2	estriol	progesterone	coprostanol (ug/g or ug/L)	cholesterol (ug/g or ug/L)
March 2006										
Primary Sludge (Unthickened)	Solid	ND	ND	ND	ND	ND	ND	ND	E 160	E 300
Thickened Wasted Activated Sludge (TWAS)	Solid	E 0.88	ND	ND	ND	ND	ND	E 8.0	E 17	E 21
Dewatering Centrate	Solid	ND	ND	ND	ND	ND	ND	E 0.35	E 14	E 16
Digested Sludge	Solid	E 3.8	ND	ND	ND	ND	E 1.1	E 64	E 140	E 33
Digested Sludge (Duplicate)	Solid	E 4.1	ND	ND	ND	ND	E 2.3	E 79	E 140	41
June 2006										
Primary Influent	Liquid	96	< 1.0	< 1.0	15	< 1.0	400	E 210	E 620	E 550
Primary Effluent	Liquid	66	< 1.0	< 1.0	< 2.6	< 1.0	350	E 310	E 820	E 690
Secondary Effluent	Liquid	< 1.0	< 1.0	< 1.0	2.2	2.8	< 1.0	< 5.0	36	41
Primary Sludge (Unthickened)	Solid	12	ND	ND	ND	ND	ND	ND	E 540	E 450
Waste Activated Sludge (Unthickened WAS)	Solid	20	ND	ND	ND	ND	ND	ND	E 110	E 170
Thickened Wasted Activated Sludge (TWAS)	Solid	11	ND	ND	ND	ND	ND	E 7.4	E 260	E 410
Thickened Wasted Activated Sludge (TWAS) (Duplicate)	Solid	10	ND	ND	ND	ND	ND	E 7.5	E 290	E 420
Thickened Wasted Activated Sludge (TWAS) (Triplicate 1)	Solid	ND	ND	ND	ND	ND	D-R	ND	25	E 49
Thickened Wasted Activated Sludge (TWAS) (Triplicate 2)	Solid	5.7	ND	ND	ND	ND	ND	E 1.7	E 83	E 180
Dewatering Centrate	Liquid	< 0.039	< 0.039	< 0.039	< 0.098	< 0.039	3.7	E 2.1	E 13	E 2.1
Dewatering Centrate	Solid	ND	ND	ND	ND	ND	10	E 3.1	E 12	E 3.9
Digested Sludge	Solid	3.7	ND	ND	ND	ND	1.4	E 22	E 590	48
TWAS Centrate	Liquid	< 0.13	0 0.15	< 0.13	< 0.33	< 0.13	0.24	< 0.67	0.41	0.64
TWAS Centrate	Solid	ND	ND	ND	ND	ND	ND	E 0.90	E 51	E 60
TWAS Centrate (Duplicate)	Solid	ND	ND	ND	ND	ND	ND	E 9.0	E 430	E 490

Notes: E = Estimated, ND = Not Detected, NA = Not Applicable, U-D = Result not reported because of sample interferences, D-R = Value is not reported because sample or analyte was ruined. Often due to matrix issues., Q-D = Result not reported because of result did not meet quality criteria, < = Denotes that the analyte was not detected; the associated parameter value is generally the reporting limit, * = Sample was not analyzed for this compound

Table 3-13. Plant D: Instantaneous Loads Results (g/day), Hormones (continued).

Sample Location	Matrix	diethylstilbestrol	cis-androsterone	epitestosterone	17-alpha-estradiol	stanolone	androstene-3,17-dione	estrone	17-beta-estradiol	testosterone	equilin
September 2006											
Primary Sludge (Unthickened)	Solid	ND	260	ND	ND	9.8	7.3	ND	5.2	ND	ND
Thickened Wasted Activated Sludge (TWAS)	Solid	ND	57	ND	ND	ND	1.5	4.1	ND	ND	ND
Digested Sludge	Solid	16	74	ND	ND	ND	140	89	44	ND	ND
TWAS Centrate	Liquid	< 0.12	< 0.12	< 0.58	< 0.12	< 0.12	0.50	1.5	< 0.12	< 0.12	< 0.58
TWAS Centrate	Solid	ND	0.94	ND	0.56	ND	0.23	0.61	1.2	ND	ND
TWAS Centrate (Duplicate)	Solid	ND	1.8	ND	ND	ND	0.37	0.50	ND	ND	ND
December 2006											
Primary Influent	Liquid	< 1.0	E 3,000	65	25	77	E 590	460	91	100	< 5.0
Primary Effluent	Liquid	< 1.0	E 2,700	68	13	92	E 630	480	67	110	< 5.2
Primary Effluent (Duplicate)	Liquid	< 1.0	E 3,400	65	13	100	E 670	450	70	110	< 5.2
Secondary Effluent	Liquid	< 1.0	< 1.0	< 5.0	< 1.0	< 1.0	< 7.5	1.9	< 1.0	< 1.0	< 5.0
Primary Sludge (Unthickened)	Solid	5.4	390	ND	2.8	11	16	30	17	7.7	ND
Primary Sludge (Unthickened) (Duplicate)	Solid	ND	200	ND	ND	5.3	10	17	6.7	6.7	ND
Waste Activated Sludge (Unthickened WAS)	Solid	D-R	D-R	D-R	D-R	D-R	D-R	D-R	D-R	D-R	D-R
Thickened Wasted Activated Sludge (TWAS)	Solid	D-R	16	ND	1.1	ND	6.1	9.9	3.2	ND	ND
Dewatering Centrate	Liquid	< 0.035	0.25	< 0.17	0.21	< 0.035	0.47	0.12	0.030	< 0.035	< 0.17
Dewatering Centrate	Solid	ND	E 28	0.79	0.29	0.58	12	12	ND	0.33	ND
Digested Sludge	Solid	4.4	E 890	ND	3.2	8.4	9.0	38	9.2	3.3	ND
Digested Sludge (Duplicate)	Solid	ND	100	ND	0.49	1.2	0.96	6.7	0.45	ND	ND
TWAS Centrate	Liquid	< 0.11	E 210	< 0.57	0.76	1.7	9.4	E 16	0.36	0.91	< 0.57

Notes: E = Estimated, ND = Not Detected, NA = Not Applicable, U-D = Result not reported because of sample interferences, D-R = Value is not reported because sample or analyte was ruined. Often due to matrix issues., Q-D = Result not reported because of result did not meet quality criteria, < = Denotes that the analyte was not detected; the associated parameter value is generally the reporting limit, * = Sample was not analyzed for this compound

Table 3-13. Plant D: Instantaneous Loads Results (g/day), Hormones (continued).

Sample Location	Matrix	11-ketotestosterone	19-norethindrone	mestranol	equilenin	17-alpha-EE2	estriol	progesterone	coprostanol (ug/g or ug/L)	cholesterol (ug/g or ug/L)
September 2006										
Primary Sludge (Unthickened)	Solid	5.6	ND	ND	ND	ND	5.2	ND	E 680	E 570
Thickened Wasted Activated Sludge (TWAS)	Solid	ND	ND	ND	ND	ND	ND	ND	E 63	12
Digested Sludge	Solid	39	ND	ND	ND	ND	6.1	E 160	E 690	E 1,100
TWAS Centrate	Liquid	< 0.12	< 0.12	< 0.12	< 0.29	< 0.12	< 0.12	< 0.58	0.11	0.37
TWAS Centrate	Solid	ND	ND	ND	ND	ND	ND	ND	E 12	13
TWAS Centrate (Duplicate)	Solid	ND	ND	ND	ND	ND	0.16	E 1.2	E 23	E 34
December 2006										
Primary Influent	Liquid	44	< 1.0	< 1.0	< 2.5	< 1.0	340	E 330	E 1,800	E 730
Primary Effluent	Liquid	53	< 1.0	< 1.0	3.8	< 1.0	350	E 360	E 850	E 430
Primary Effluent (Duplicate)	Liquid	39	< 1.0	< 1.0	< 2.6	< 1.0	310	E 570	E 1,000	E 350
Secondary Effluent	Liquid	< 1.0	< 1.0	< 1.0	4.6	< 1.0	< 1.0	< 5.0	36	35
Primary Sludge (Unthickened)	Solid	ND	ND	ND	ND	ND	ND	ND	E 430	E 510
Primary Sludge (Unthickened) (Duplicate)	Solid	ND	ND	ND	ND	ND	ND	ND	E 240	E 260
Waste Activated Sludge (Unthickened WAS)	Solid	D-R	D-R	D-R	D-R	D-R	D-R	D-R	D-R	D-R
Thickened Wasted Activated Sludge (TWAS)	Solid	ND	ND	ND	ND	ND	ND	E 4.1	E 69	E 110
Dewatering Centrate	Liquid	< 0.035	< 0.035	< 0.035	< 0.087	< 0.035	< 0.035	< 0.17	0.13	0.21
Dewatering Centrate	Solid	ND	ND	ND	ND	ND	1.1	E 64	E 340	E 94
Digested Sludge	Solid	ND	ND	ND	ND	ND	9.7	E 110	E 2,000	E 380
Digested Sludge (Duplicate)	Solid	2.0	ND	ND	ND	ND	1.6	E 10	E 190	E 31
TWAS Centrate	Liquid	0.37	< 0.11	< 0.11	< 0.28	< 0.11	9.7	E 11	E 74	E 12

Notes: E = Estimated, ND = Not Detected, NA = Not Applicable, U-D = Result not reported because of sample interferences, D-R = Value is not reported because sample or analyte was ruined. Often due to matrix issues., Q-D = Result not reported because of result did not meet quality criteria, < = Denotes that the analyte was not detected; the associated parameter value is generally the reporting limit, * = Sample was not analyzed for this compound

Table 3-14. Plant D: Instantaneous Loads Results (g/day), Alkylphenolic Compounds.

Sample Location	Matrix		4-Nonylphenol		4-Nonylphenol monoethoxylates		4-Nonylphenol diethoxylates		4-tert-Octylphenol		4-Octylphenol monoethoxylates		4-Octylphenol diethoxylates		5-Methyl-1H-benzotriazole		Bisphenol A		beta-Sitosterol		1,4-Dichlorobenzene		4-Cumylphenol
March 2006																							
Dewatering Centrate	Liquid	U-D		E	150	E	240	E	30	<	8.7	E	27	U-D		E	17	E	62	E	5.6	<	8.7
Dewatering Centrate (Duplicate)	Liquid	<	44	<	17	<	44	<	8.7	<	8.7	<	8.7	<	17	U-D		<	17	<	4.4	<	8.7
June 2006																							
Primary Influent	Liquid	E	7,900	E	6,900	E	16,000	E	890	E	1,100	E	2,200	<	2,500	U-D		E	7,000	E	1,300	<	1,300
Primary Effluent	Liquid	E	8,200	E	6,500	E	14,000	E	1,100	E	1,300	E	2,000	<	2,600	U-D		E	4,400	E	1,200	<	1,300
Secondary Effluent	Liquid	E	3,600	E	18,000	E	19,000	E	330	E	3,100	E	2,900	E	4,000	E	140	E	710	E	1,300	<	1,200
Primary Sludge (Unthickened)	Solid	E	75,000	E	180,000	E	79,000		4,500	E	7,100	<	3,800	*		E	3,200	E	77,000	E	690	<	3,800
Waste Activated Sludge (Unthickened WAS)	Solid	E	19,000	E	110,000	E	54,000	E	570	E	5,200	E	2,200	*		U-D		E	29,000	<	5,300	<	5,300
Thickened Wasted Activated Sludge (TWAS)	Solid	E	9,400	E	25,000	E	15,000	E	260	E	1,400	E	1,000	*		E	420	E	2,800	E	85	<	260
TWAS Centrate	Liquid	E	190		220	E	320	<	33	E	44	E	36	<	67	U-D		<	67	E	29	<	33
Digested Sludge	Solid	E	150,000	E	23,000	E	15,000		5,500	E	810	<	2,100	*		E	2,100	E	31,000	<	2,100	<	2,100
Dewatering Centrate	Liquid	E	400	E	140	E	260		57	<	9.8	E	26	E	600,000	U-D		E	110	E	7.4	<	49

Notes: E = Estimated, ND = Not Detected, NA = Not Applicable, U-D = Result not reported because of sample interferences, Q-D = Result not reported because of result did not meet quality criteria, < = Denotes that the analyte was not detected; the associated parameter value is generally the reporting limit, * = Sample was not analyzed for this compound

Table 3-14. Plant D: Instantaneous Loads Results (g/day), Alkylphenolic Compounds (continued).

Sample Location	Matrix	4-Nonylphenol	4-Nonylphenol monoethoxylates	4-Nonylphenol diethoxylates	4-tert-Octylphenol	4-Octylphenol monoethoxylates	4-Octylphenol diethoxylates	5-Methyl-1H-benzotriazole	Bisphenol A	beta-Sitosterol	1,4-Dichlorobenzene	4-Cumylphenol
September 2006												
Primary Sludge (Unthickened)	Solid	E 110,000	E 270,000	E 110,000	8,000	E 14,000	E 9,300	*	U-D	E 190,000	< 2,800	< 2,800
Thickened Wasted Activated Sludge (TWAS)	Solid	E 14,000	E 25,000	E 20,000	E 590	E 2,100	E 1,700	*	U-D	E 28,000	< 660	< 660
TWAS Centrate	Liquid	< 140	< 140	< 140	< 140	< 140	< 140	< 140	U-D	< 140	< 140	< 140
Digested Sludge	Solid	E 8,400,000	E 1,500,000	E 2,400,000	E 460,000	< 2,500,000	< 490,000	*	U-D	E 2,500,000	< 490,000	< 490,000
December 2006												
Primary Influent	Liquid	E 93,000	E 150,000	E 250,000	17,000	E 19,000	E 29,000	< 51,000	U-D	E 150,000	E 8,000	< 2,500
Primary Effluent	Liquid	E 130,000	E 78,000	E 190,000	19,000	E 17,000	E 31,000	< 52,000	U-D	E 240,000	E 7,300	< 2,600
Secondary Effluent	Liquid	E 2,900	E 9,200	E 14,000	180	E 1,100	E 1,800	E 3,400	E 260	E 2,100	E 750	< 170
Primary Sludge (Unthickened)	Solid	E 95,000	E 95,000	E 10,000	E 5,300	E 11,000	< 9,800	*	< 9,800	E 160,000	< 9,800	< 9,800
Waste Activated Sludge (Unthickened WAS)	Solid	E 14,000	E 14,000	E 33,000	< 3,000	E 24,000	E 3,600	*	< 3,000	E 32,000	< 3,000	< 3,000
Thickened Wasted Activated Sludge (TWAS)	Solid	E 7,800	E 13,000	E 11,000	E 180	E 630	E 630	*	E 180	E 12,000	E 42	< 250
TWAS Centrate	Liquid	E 170	E 500	E 790	E 12	E 72	E 94	E 510	E 8.8	< 57	E 14	< 57
Digested Sludge	Solid	E 4,100,000	E 4,100,000	E 21,000	14,000	E 2,400	< 2,500	*	E 2,500	E 81,000	< 2,500	< 2,500
Dewatering Centrate	Liquid	E 460	E 110	E 130	< 17	< 17	< 17	E 320,000	E 15	< 17	E 4.2	< 17

Notes: E = Estimated, ND = Not Detected, NA = Not Applicable, U-D = Result not reported because of sample interferences, Q-D = Result not reported because of result did not meet quality criteria, < = Denotes that the analyte was not detected; the associated parameter value is generally the reporting limit, * = Sample was not analyzed for this compound

Table 3-15. Plant D: Instantaneous Loads Results, YES Bioassay.

Sample Location	Eluent Fraction	Instantaneous Load (g/day, EE2 Eqs)			
		Mar-06	Jun-06	Sep-06	Dec-06
Primary Influent (L)	20 50 80	NA	15	NA	29
Primary Influent (L) (Duplicate)	20 50 80	NA	31	NA	NA
Primary Influent (S)	20 50 80	NA	T	NA	1.5
Primary Effluent (L)	20 50 80	NA	23	NA	4.3
Primary Effluent (S)	20 50 80	NA	NA	NA	19
Secondary Effluent (L)	20 50 80	NA	2.3	NA	0.17
Secondary Effluent (L) (Duplicate)	20 50 80	NA	1.4	NA	NA
Primary Sludge (Unthickened)	20 50 80	2.2 T*	2.1	2.1	0.78 T*

Notes: NR = no estrogenic response from sample, T = sample contained toxicity (no estrogenic response was observed), T* = sample contained toxicity (estrogenic response was also observed but was not quantified due to presence of toxicity), NA = not analyzed

Table 3-15. Plant D: Instantaneous Loads Results, YES Bioassay (continued).

Sample Location	Eluent Fraction	Instantaneous Load (g/day, EE2 Eqs)			
		Mar-06	Jun-06	Sep-06	Dec-06
Waste Activated Sludge (Unthickened)	20 50 80	NA	9.4	NA	0.18
Thickened Waste Activated Sludge (TWAS)	20 50 80	2.9	2.9	1.9	0.11
Centrate Recycle Stream from Dewatering Process (L)	20 50 80	0.042 T*	0.075	NA	0.0071
Centrate Recycle Stream from Dewatering Process (L) (Blank)	20 50 80	0.0012	NA	NA	NA
Centrate Recycle Stream from Dewatering Process (S)	20 50 80	0.64	0.24	NA	NR
Digested Sludge	20 50 80	10	4.4 T*	19	6.3
Digested Sludge (Duplicate)	20 50 80	1.1 T*	NA	NA	NA
TWAS Centrate (L)	20 50 80	NA	0.34	0.40	0.064
TWAS Centrate (S)	20 50 80	NA	0.23	0.096	0.0024 T*

Notes: NR = no estrogenic response from sample, T = sample contained toxicity (no estrogenic response was observed), T* = sample contained toxicity (estrogenic response was also observed but was not quantified due to presence of toxicity), NA = not analyzed

3.6.2 Chemical Analysis: Data Reduction Results and Discussion

3.6.2.1 Steroids

Compared to the other plant where liquid samples were collected (Plant B), Plant D does a more effective job at transformation of steroids, and there is no seasonal variation (Figures 3-17 and 3-18). We speculate that this may be because of the location of Plant D in a warmer climate than Plant B, or it could be the result of some other difference with the treatment process; additional high frequency chemical sampling and process monitoring are necessary to determine whether temperature or other causes produce the observed differences in steroid transformation. The instantaneous load of steroidal estrogens entering the plant averages over 700 g/day and is composed primarily of E1, E2, and E3, although 17α-estradiol and equilenin are also present. The outgoing load in the secondary effluent is less than 10 g/day and the three major constituents listed above are all removed (transformed) with greater than 95% efficiency. The extent of transformation was calculated by accounting for total outgoing load in both the secondary effluent and the TWAS. Once again, the concentrations in the unthickened stream were highly variable due to the small mass of solids that were extracted and it was determined that assessing load downstream of the thickening process was most reliable. It is interesting to note that the synthetic hormone EE2 is present in the secondary effluent but not in the primary influent; this is likely an anomaly due to its presence at very close to the analytical detection limit and the presence of greater amounts of interfering organic matter in the untreated stream. Six of the androgens were removed that were present in the primary effluent and they too were consistently removed with greater than 95% efficiency. Due to the high level of hormone removal during secondary treatment and the relatively low solids content of the digestor feeds (primary sludge, TWAS), the effectiveness of the anaerobic digestor is difficult to evaluate with respect to the hormones. A number of compounds appear to increase during the course of digestion but this is likely an anomaly due to low detection frequency in the primary sludge and TWAS, and/or variability in results. It is unlikely that there is precursor material to form these hormones after aerobic treatment has occurred, but it is interesting to note that the digester feed is a combination of primary and secondary waste sludge. The load of hormones in the primary stream that has not undergone biological treatment is often somewhat higher than in the TWAS stream. For those compounds (e.g., E1, E2) that can be formed by interconversion of other steroids, the primary waste sludge is a potential source of material. Also, it is possible that sulfate and glucoronide conjugates of the steroids are present in the untreated primary sludge and cleaved during anaerobic digestion. This could also explain an increase in load of hormones during digestion (Figure 3-19). Many studies have shown that most conjugated hormones are cleaved prior to arrival at the WWTP and it is unknown why Plant D would have apparently greater incidence of conjugates than the other plants.

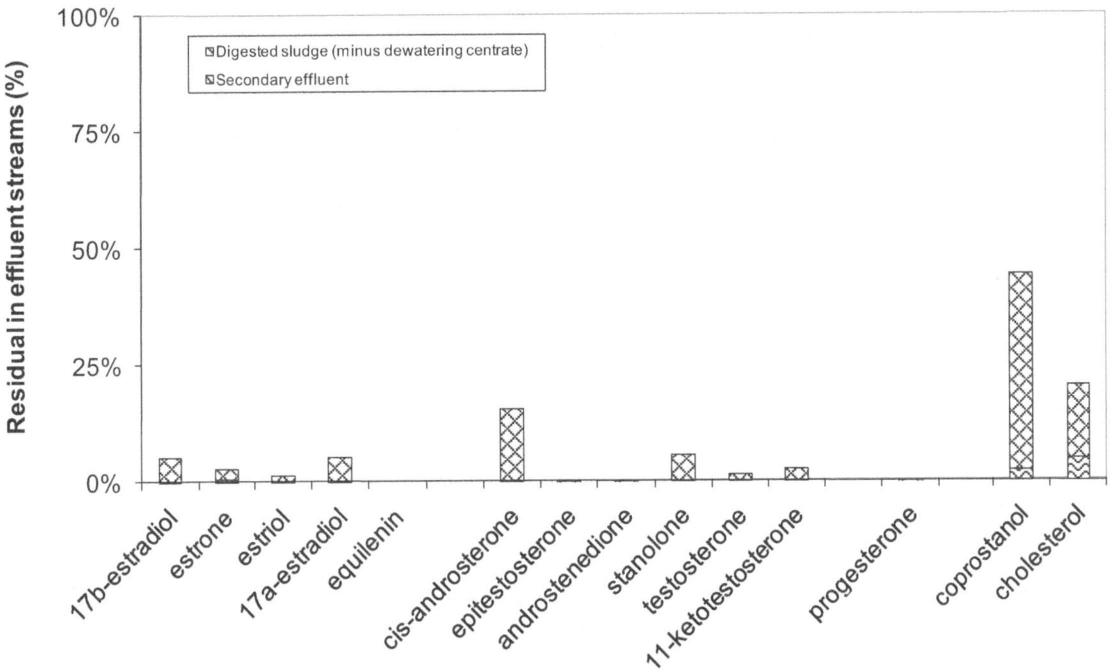

Figure 3-17. Plant D: Hormone Removal (December 2006).

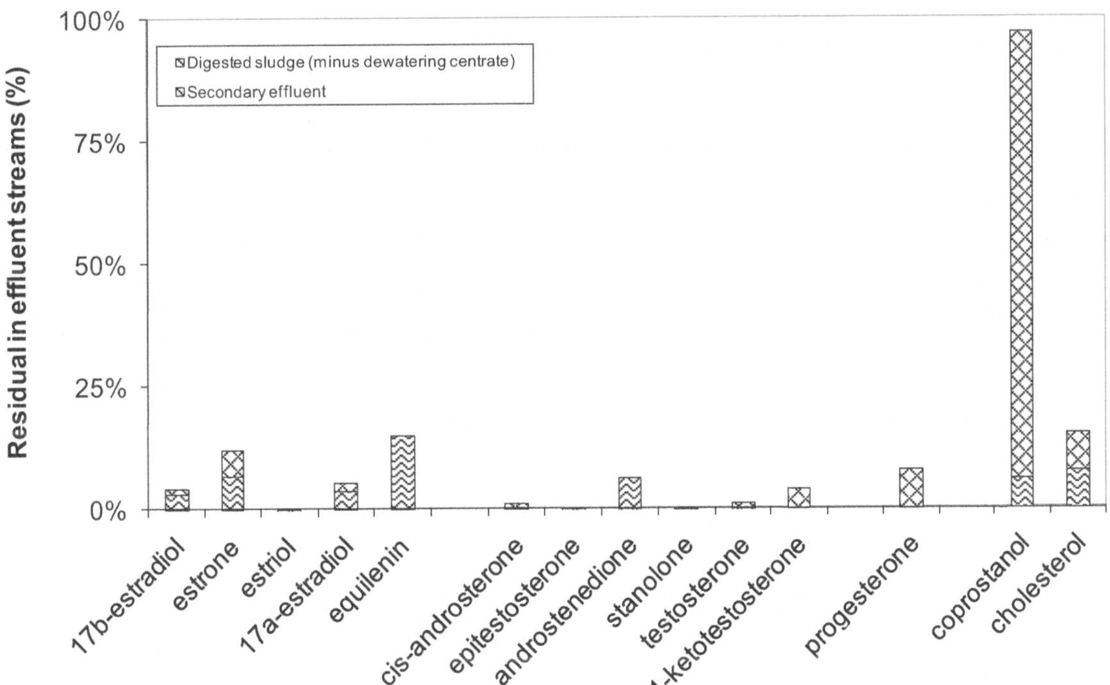

Figure 3-18. Plant D: Hormone Removal (June 2006).

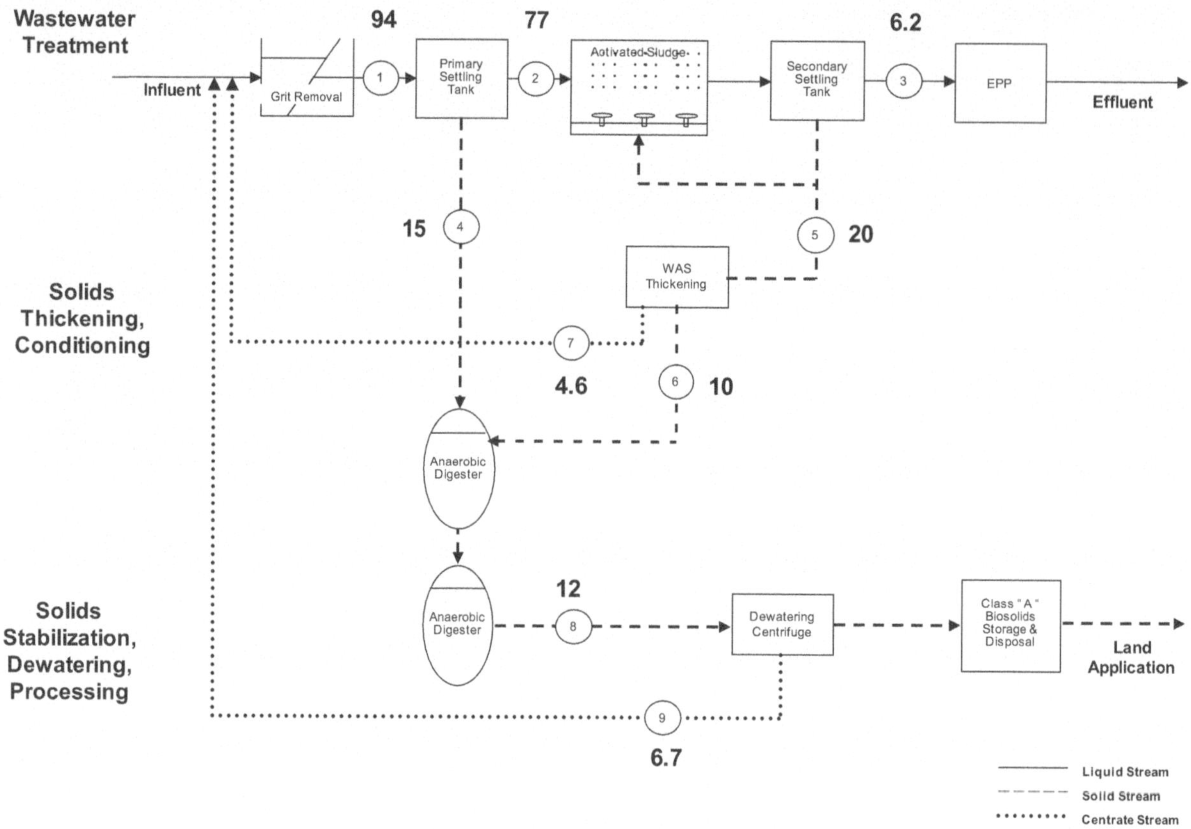

Figure 3-19. Estrone Flux (g/day) Through Plant D (June 2006).

3.6.2.2 Non-Steroidal Estrogenic Compounds

As was observed in Plant B, aerobic treatment at Plant D does not effectively remove short chain alkylphenols. Once again, this is most likely to be the result of biotransformation of long chain alkylphenols that are not detected using the current methodology. While certain compounds can have less load in the secondary effluent than in the primary influent (e.g. NP, NP1EO, NP2EO in December 2006) the decrease in load is due mainly to partitioning to solid material. During the same sampling period, when load in primary waste sludge and TWAS are taken into account, load of NP and NP1EO actually increase slightly while NP2EO load decreases somewhat. The effect is even more pronounced in July of 2006, when all three NPEs increase by more than a factor of 10. Similar trends are also evident with the OPEOs. When we consider the effects of anaerobic digestion on APEOs, the effects are once again mixed. In the winter, both NP and NP1EO increase substantially (by a factor of 25 or more) during digestion while NP2EO decreases by about 50%. In the summer, only NP increases during digestion and only by a factor of 2, while both NP1EO and NP2EO decrease by more than 80%.

3.6.3 Biological Analysis: Data Reduction Results, Model of Concentration Addition, and Discussion

The average (n = 4) daily mass fluxes of estrogenicity (moles of EE2 equivalents per day), based on the Model of Concentration Addition, were determined in Plant D for primary influent, primary effluent, secondary effluent, primary sludge, waste-activated sludge, and the combined solids after thermophilic anaerobic digestion (Figure 3-20). 17β-E2 and E1 were the most prominent contributors in primary influent, primary effluent and secondary effluent. Estrogenicity was substantially reduced during secondary treatment at Plant D. The amount of estrogenicity remaining in secondary effluent represented about 3% of the total estrogenicity present in primary influent. Steroidal hormones accounted for the majority of estrogenicity in primary and secondary effluents.

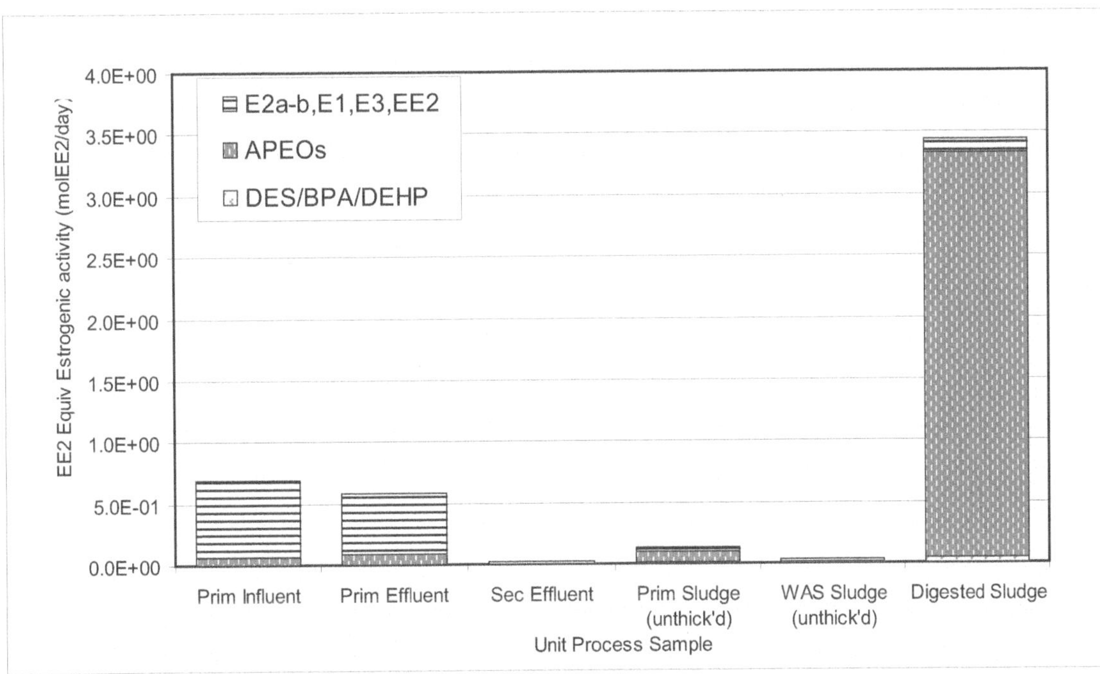

Figure 3-20. Average Daily Estrogenicity Mass Fluxes (mol EE2-equivalents/day) at Plant D. (Based on the Model of Concentration Addition)

The estrogenic contributions of nonylphenol and octylphenol in primary and waste-activated sludge were approximately equal to that from the steroidal hormones. Comparing the mass flux of estrogenicity in the primary sludge plus the waste-activated sludge versus the flux in the anaerobically digested solids indicates there was a very substantial net production of estrogenicity in the solids as a consequence of thermophilic anaerobic digestion. The increase in estrogenicity flux after thermophilic anaerobic digestion was almost entirely due to a greater contribution by APEOs, particularly nonylphenol, which is created during the breakdown of aklylphenol polyethoxylates under anaerobic conditions and has a much higher estrogenic potency than the parent alkylphenol polyethoxylates.

Average (n = 4) daily mass fluxes of estrogenic activity based on the YES bioassay measurements are shown in Figure 3-21. The YES bioassay results exhibit generally consistent trends with the Model of Concentration Addition results. The chemical and bioassay measurements both reveal that there was a greater mass flux of estrogenicity discharged from this facility in the solids than was discharged in the secondary effluent, consistent with the finding at Plant B.

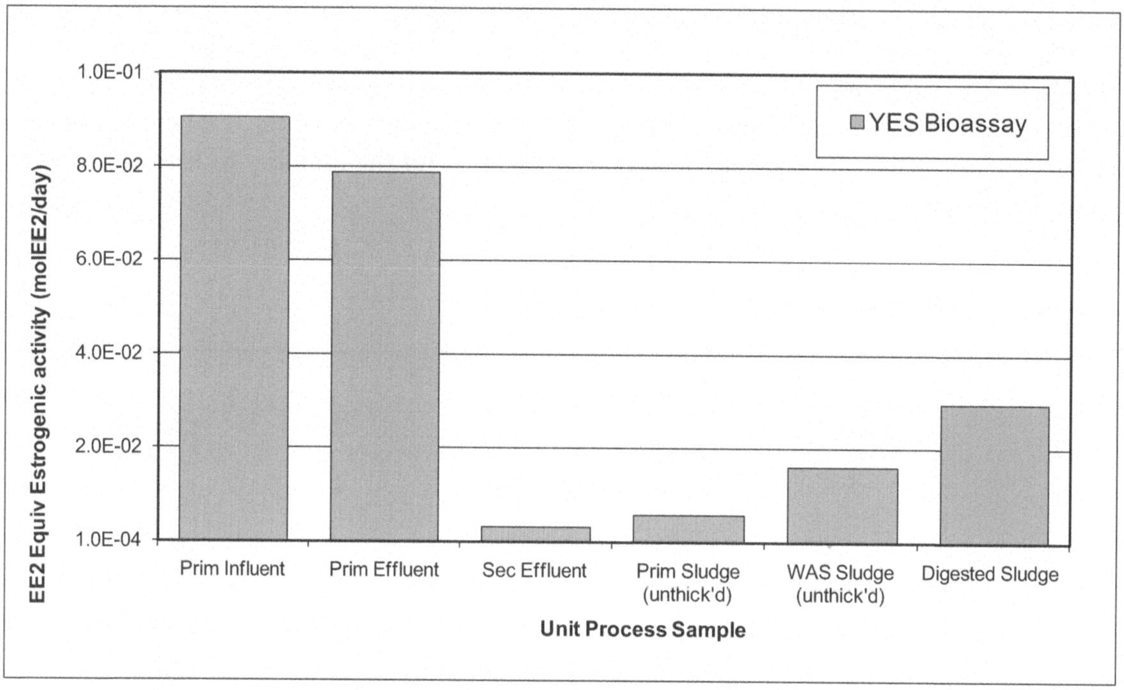

Figure 3-21. Average Daily Mass Fluxes (mol EE2-equivalents/day) of Estrogenic Activity at Plant D. (Based on the YES Bioassay Measurements)

The magnitude of estrogenic activity daily mass fluxes (based on the YES bioassay measurements) at various sampling points in Plant D are shown in Figure 3-22 as percentages of the mass fluxes determined from the Model of Concentration Addition. The YES-based mass fluxes in liquid-stream samples varied from 13% (primary influent and effluent) to 20% (secondary effluent) with the waste-activated sludge and post-digested solids YES responses of 51% and 2%, respectively, relative to the corresponding mass fluxes calculated using the Model of Concentration Addition. This result is consistent with what was observed at the other Plants in the study. Possible mechanisms accounting for the lower response from the YES bioassay are discussed in Section 3.4.3 (Plant B).

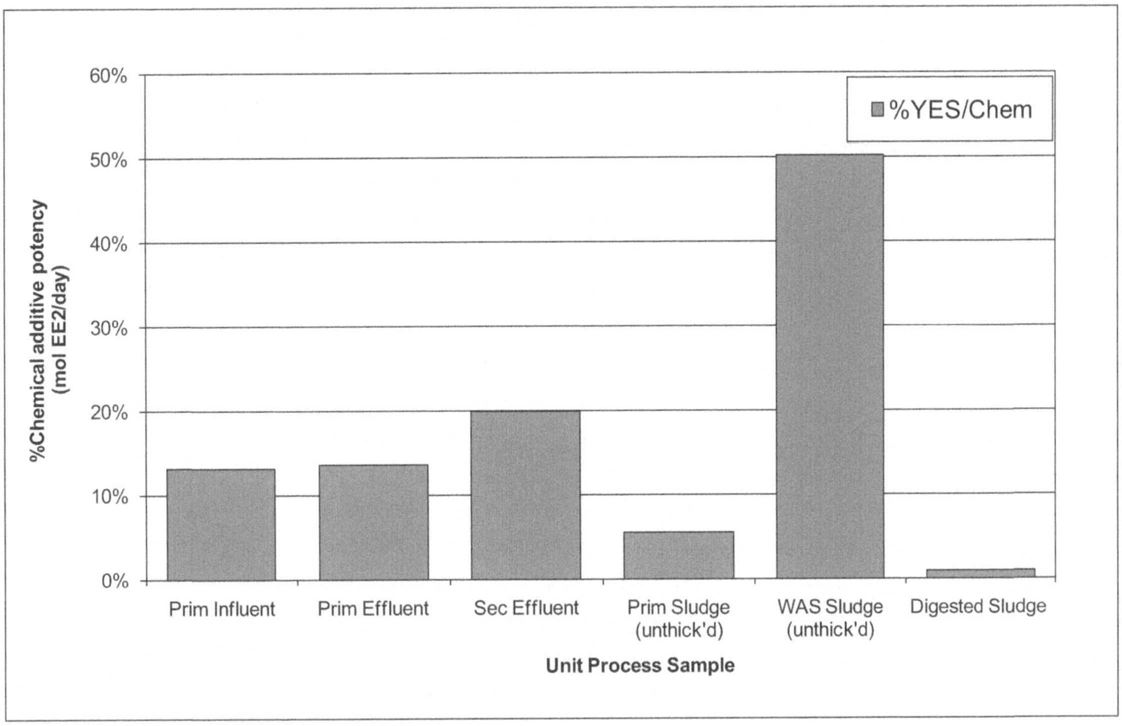

Figure 3-22. YES Bioassay Estrogenic Response as a% of Response Calculated Using the Model of\ Concentration Addition. (Calculated Using Data Shown in Figures 3-20 and 3-21).

A summary of the liquid and solid hydraulic fluxes at Plant D during all four sampling periods is provided in Table 3-15. The summary reveals an average decrease in liquid flow across the entire plant of 0.16% ± 0.08%. The solid fluxes through the plant decreased an average of 44% ± 5%. The liquid flows around the activated sludge process decreased by 1.5% ± 0.05% where as the solids flows increased 23% ± 12%. Two of the post-digestion streams were not sampled: a liquid recycle stream (DSF recycle) that returned to the head of the plant and the fraction of the digested solids that were disposed to a landfill. The DSF recycle liquid flow was equal to that leaving in digested solids (sample point 8). The solids delivered to the landfill constituted about 4% of the total solids flux at sample point 8. After subtracting the recycle line flow from the thermophilic anaerobic digestion control volume, the liquid flow across the digester increased by 37% ± 14% and the solids flux decreased by 53% ± 5%. The instantaneous loads analyses were based on the average hydraulic flux balances in Table 3-16.

Table 3-16. Summary of Hydraulic Balances for Liquid and Solids Flows at Plant D for Each of the Four Sampling Periods.

Unit Process	Date	Liquid Flux	Solids Mass Flux
Activated sludge	March 2006	1.5%	-36%
	June 2006	1.4%	-29%
	September 2006	1.4%	-11%
	December 2006	1.5%	-16%
	Average	*1.5%*	*-23%*
	Standard Deviation	*0.05%*	*12%*
Thermophilic Anaerobic Digester	March 2006	-26%	59%
	June 2006	-35%	56%
	September 2006	-30%	50%
	December 2006	-58%	50%
	Average	*-37%*	*54%*
	Standard Deviation	*14%*	*4.5%*
Entire Plant	March 2006	-0.25%	50%
	June 2006	-0.11%	39%
	September 2006	-0.05%	42%
	December 2006	-0.24%	44%
	Average	*-0.16%*	*44%*
	Standard Deviation	*0.10%*	*4.7%*

Note: Positive values indicate a net decrease in hydraulic flux through the unit operation; negative values indicate an increase

Liquid stream flows and waste activated sludge were sampled and analyzed twice (June 2006 and December 2006); the December waste activated sludge hormone and DES results were not reported because of matrix interferences. Bisphenol A was not reported in influent or primary effluent also because of matrix interferences. DEHP was reported only for the solid waste streams.

Instantaneous loads analyses were conducted around the activated sludge process, the thermophilic anaerobic digester, as well as the entire plant. Steroidal hormones contributed more than 75% of the total estrogenicity in influent, primary effluent, and secondary effluent, based on the Model of Concentration Addition (Figures 3-23 and 3-24). About 50% of the estrogenicity mass flux in influent and primary effluent was attributed to estrone. The estrogenic hormones as a group decrease by 94% during activated sludge treatment at Plant D with estrone decreasing by two orders of magnitude. Estrone comprised 29% of the remaining estrogenicity flux in secondary effluent, similar to the contribution provided by EE2 (28%). The combined contributions of 17α- and 17β-estradiol in the liquid-stream estrogenicity mass fluxes decreased from approximately 40% to 20% during secondary treatment.

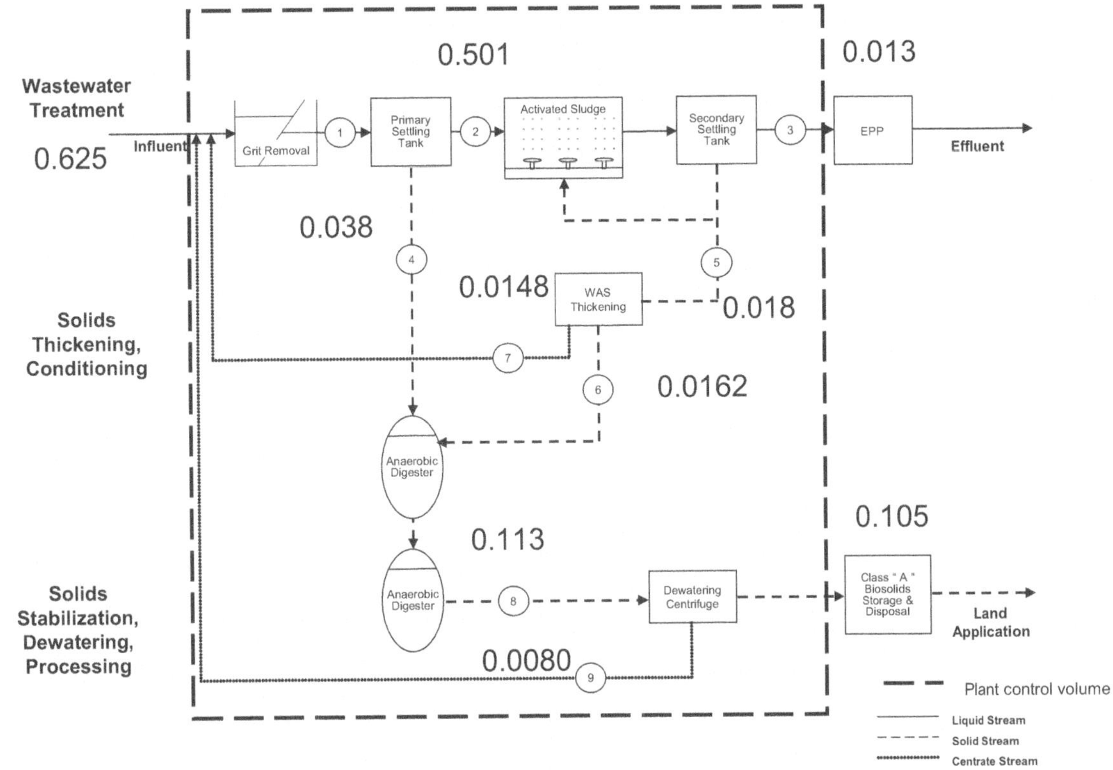

Figure 3-23. Daily Estrogenicity Mass Fluxes (mol EE2 equivalents/day) Due to Estrogenic Hormones (E2-α,E2-β,E1,E3, EE2) Through Unit Treatment Processes at Plant D. (Based on the Model of Concentration Addition)

During thermophilic anaerobic digestion, there was a 109% increase in the estrogenicity mass flux from steroidal hormones (Figure 3-24). The mass flux of steroidal hormones in waste activated sludge was similar to that in secondary effluent. Estrone comprised 37% of the estrogenicity contribution in waste activated sludge and 17α- plus 17β-estradiol provided 19%.

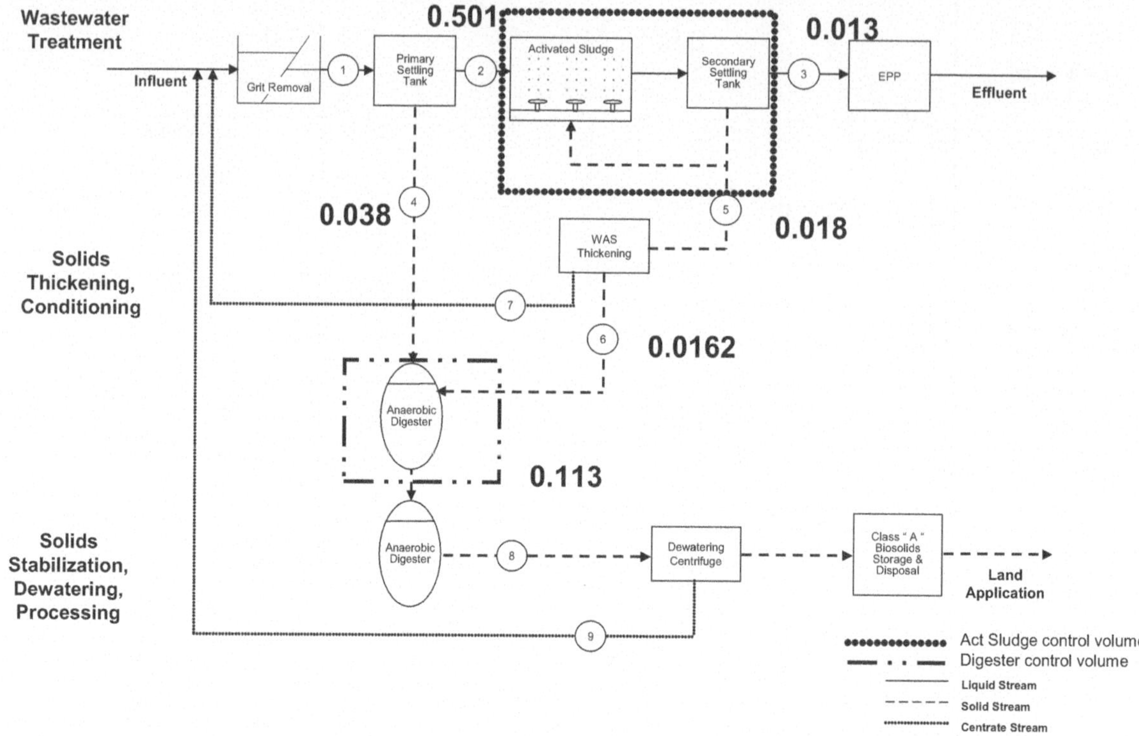

Figure 3-24. Daily Estrogenicity Mass Fluxes (mol EE2 equivalents/day) by Steroidal Hormones (E2-α,E2-β,E1,E3, and EE2) During Activated Sludge Treatment and Thermophilic Anaerobic Digestion at Plant D. (Based on the Model of Concentration Addition)

Nonylphenol and 4-*tert*-octylphenol provided approximately 5% each of the total estrogenicity mass flux in primary influent and primary effluent, while shorter chain ethoxylates of NP and OP were less than 0.5%. The estrogenicity mass flux of APEOs decreased by 80% during activated sludge secondary treatment at Plant D (Figures 3-25 and 3-26). The contribution of nonylphenol to the total remaining estrogenicity mass flux after secondary treatment increased to 12%. Nonylphenol comprised 34% of the estrogenicity mass flux in waste activated sludge.

The mass balance results around secondary treatment for the estrogenic compounds DES/ BPA/ DEHP show an approximate 59% reduction (Figure 3-27), even though the findings' strength is affected by a dearth of data for BPA and DEHP in the primary clarifier effluent and secondary effluent. There was a net increase across the plant of 5,020% in estrogenicity mass flux for APEOs and an increase of >1,330% for the DES/BPA/DEHP group.

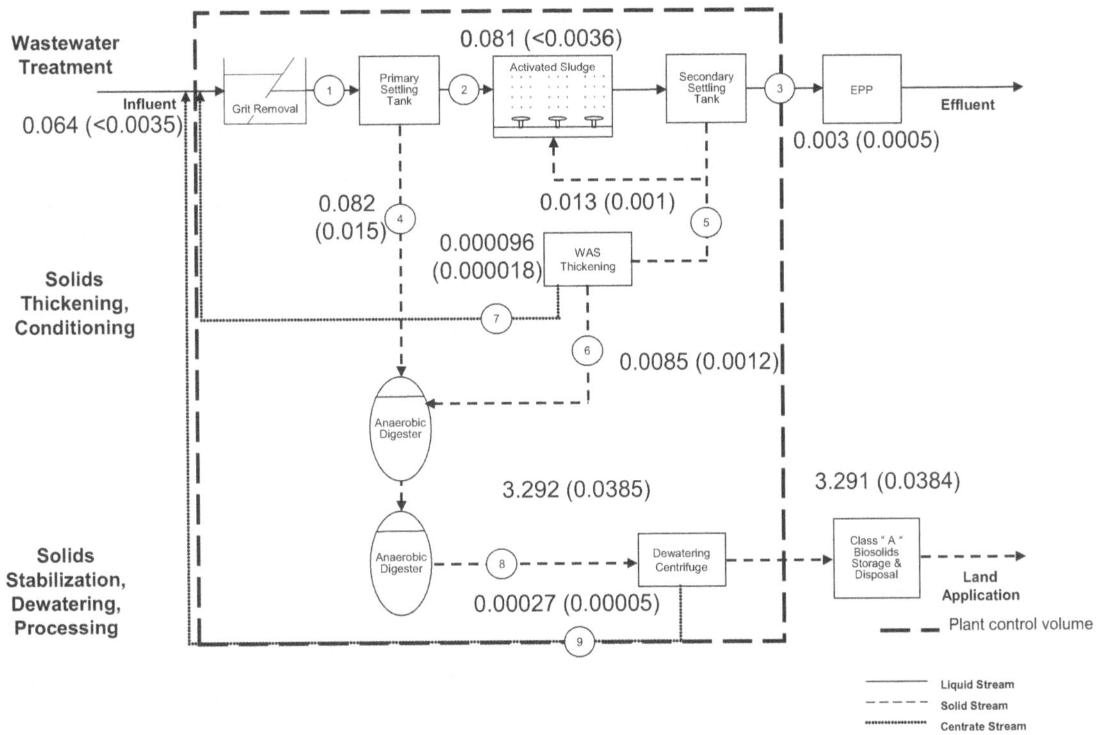

Figure 3-25. Daily Estrogenicity Mass Fluxes (mol EE2 equivalents/day) by Total APEOs and DES/BPA/DEHP (parentheses) Through Unit Treatment Processes at Plant D. (Based on the Model of Concentration Addition)

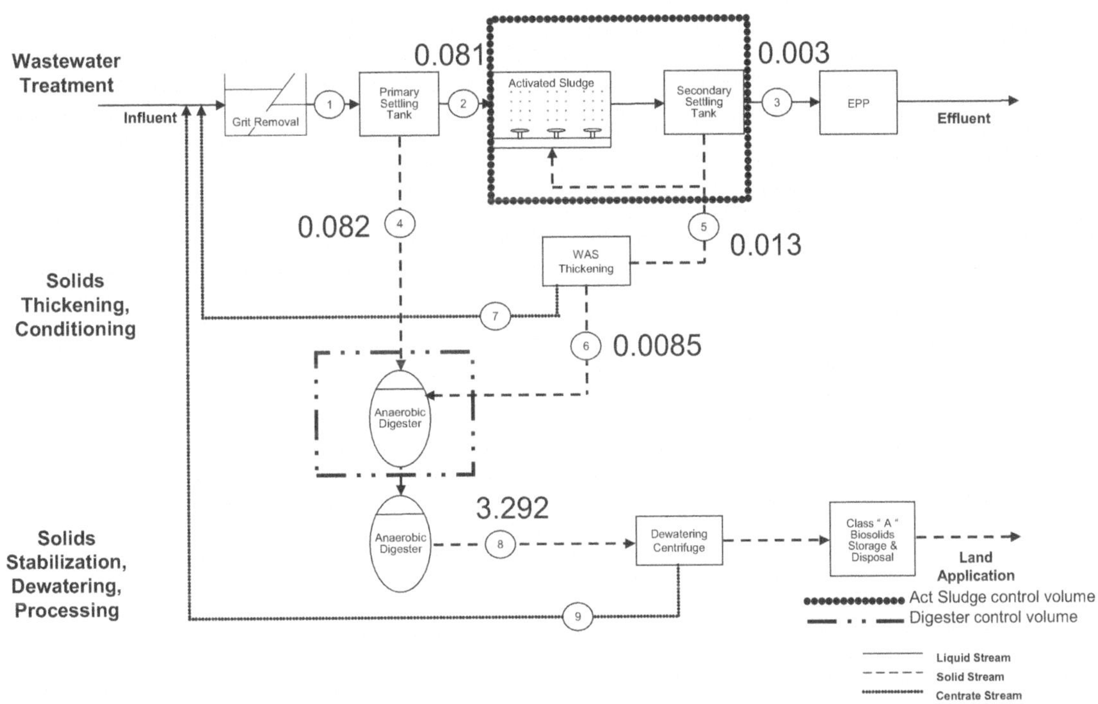

Figure 3-26. Daily Estrogenicity Mass Fluxes (mol EE2 equivalents/day) by Total APEOs during Activated Sludge Treatment and Thermophilic Anaerobic Digestion at Plant D. (Based on the Model of Concentration Addition)

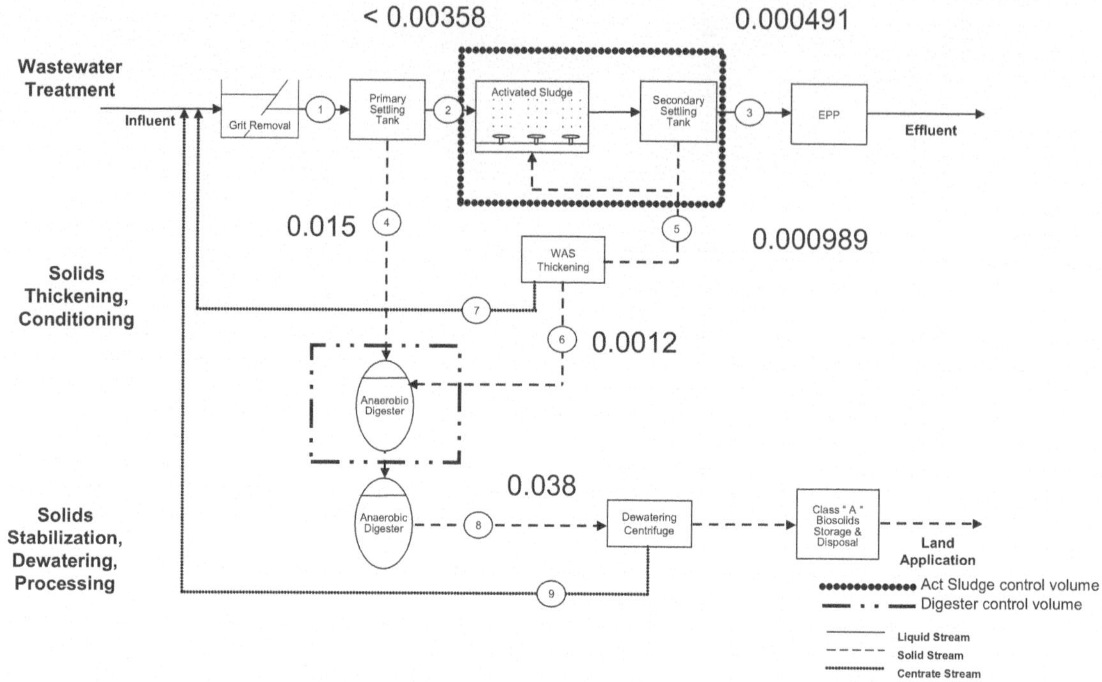

Figure 3-27. Daily Estrogenicity Mass Flux (mol EE2 equivalents/day) by DES/BPA/DEHP during Activated Sludge Treatment and Thermophilic Anaerobic Digestion at Plant D. (Based on the Model of Concentration Addition)

Daily mass fluxes of estrogenic activity, based on the YES bioassay measurements, are shown in Figures 3-28 and 3-29. There was a 65% net reduction in estrogenic activity mass flux across the entire plant. Considering unit processes separately, the activated sludge secondary treatment process provided a 75% reduction and the thermophilic anaerobic digestion process a 135% increase in estrogenic activity mass flux (Figure 3-29).

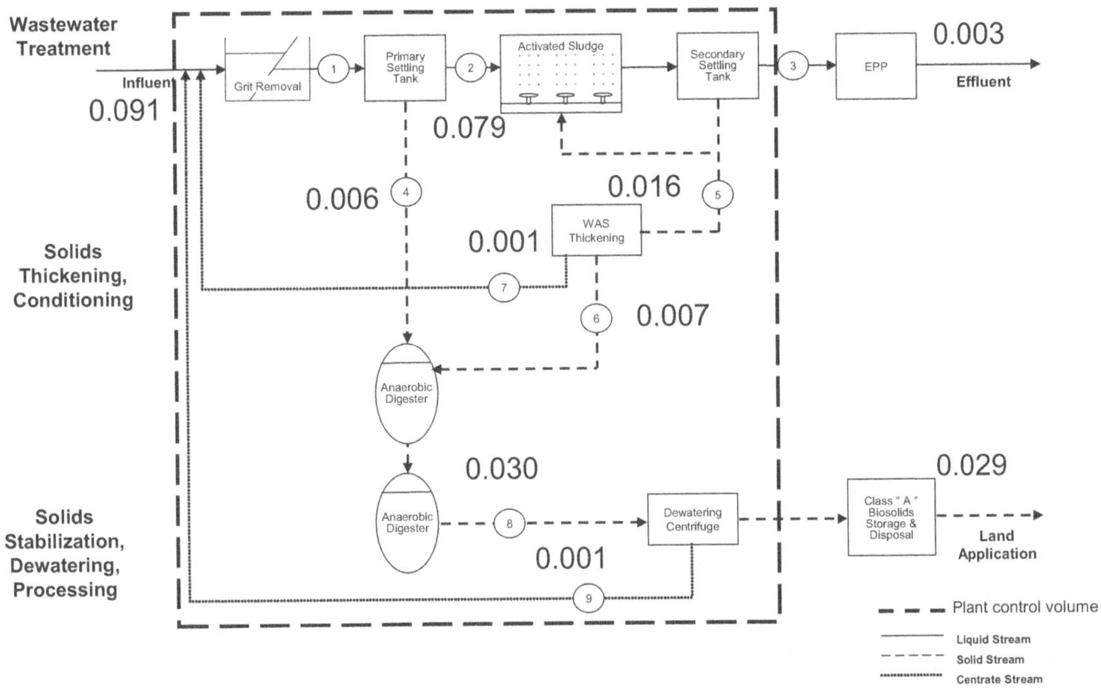

Figure 3-28. Daily Estrogenic Activity Mass Flux (mol EE2 equivalents/day) Through Unit Treatment Processes at Plant D. (Based YES Bioassay Measurements)

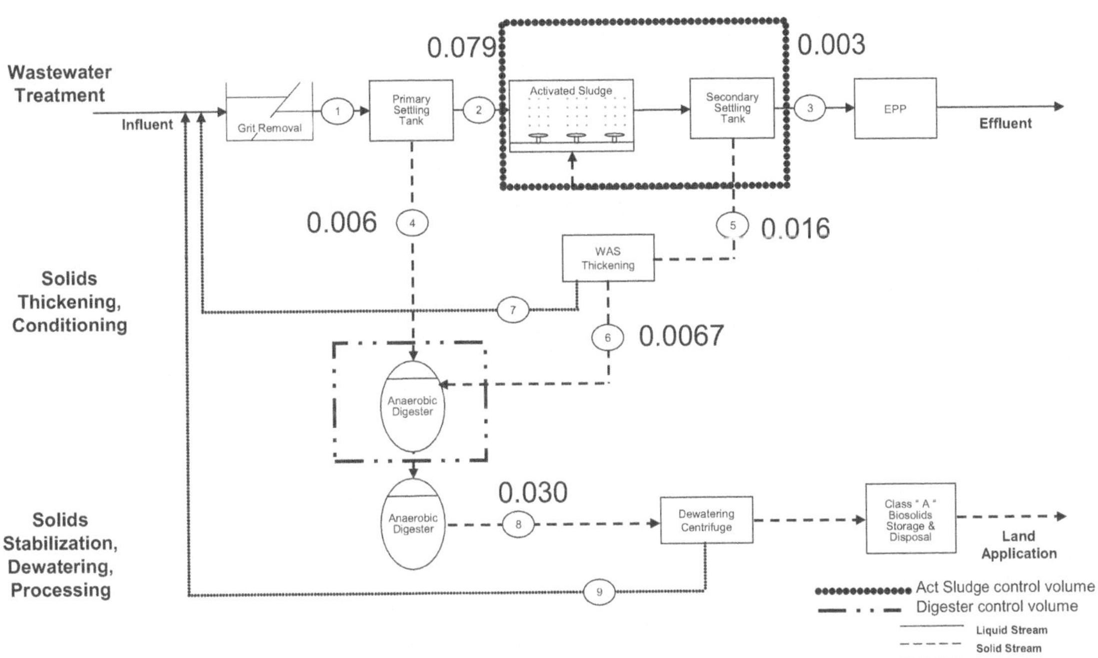

Figure 3-29. Daily Estrogenic Activity Mass Flux (mol EE2 equivalents/day) during Activated Sludge Treatment and Thermophilic Anaerobic Digestion at Plant D. (Based on the YES Bioassay Measurements)

The YES and KBluc bioassays were applied in parallel for the June 2006 sample set from Plant D. Results were used to develop parallel sets of estrogenic activity mass fluxes through unit treatment processes at Plant D (Figures 3-30 and 3-31, respectively). Comparison of the figures reveals both similarities and differences between the YES and KBluc bioassay measurements for liquid-stream and solid-stream treatment processes at Plant D.

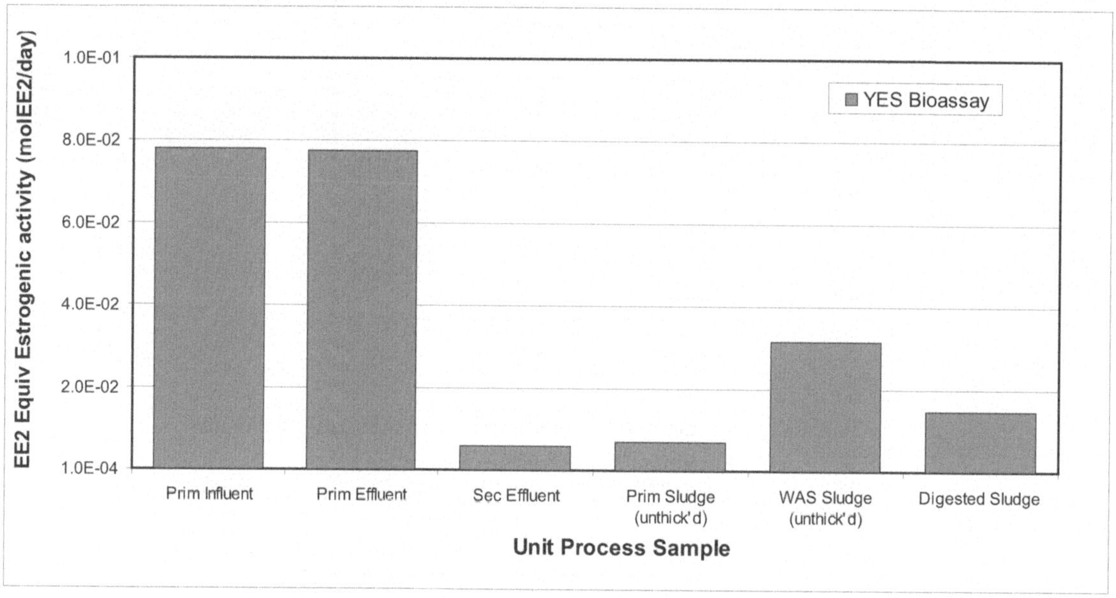

Figure 3-30. Daily Mass Flux of Estrogenic Activity at Plant D. (Based on YES Bioassay Measurements - June 2006 Data Only)

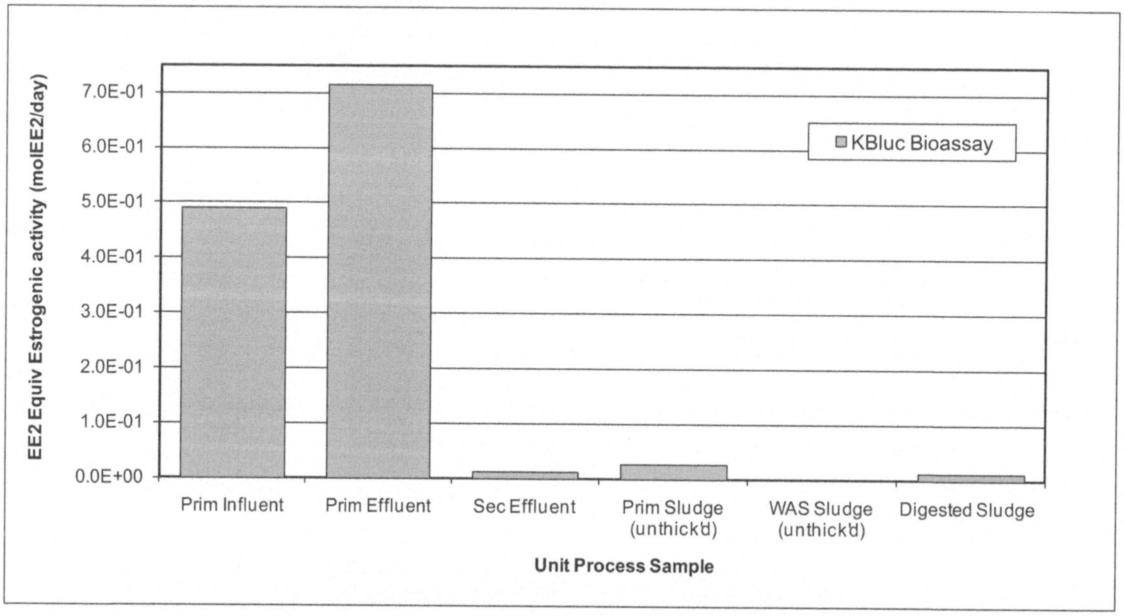

Figure 3-31. Daily Mass Flux of Estrogenic Activity at Plant D. (Based on the T47D-KBluc Bioassay Measurements – June 2006 Data Only)

In addition, estrogenicity mass fluxes were calculated as the sum of individual compound concentrations multiplied by their YES-based (Table 3-1) and KBluc-based (Table 3-1) EE2 potency factors. The two resultant Models of Concentration Addition (Figures 3-32 and 3-33, respectively) also reveal similarities and differences of the EE2 potency factors derived from the two bioassay methods.

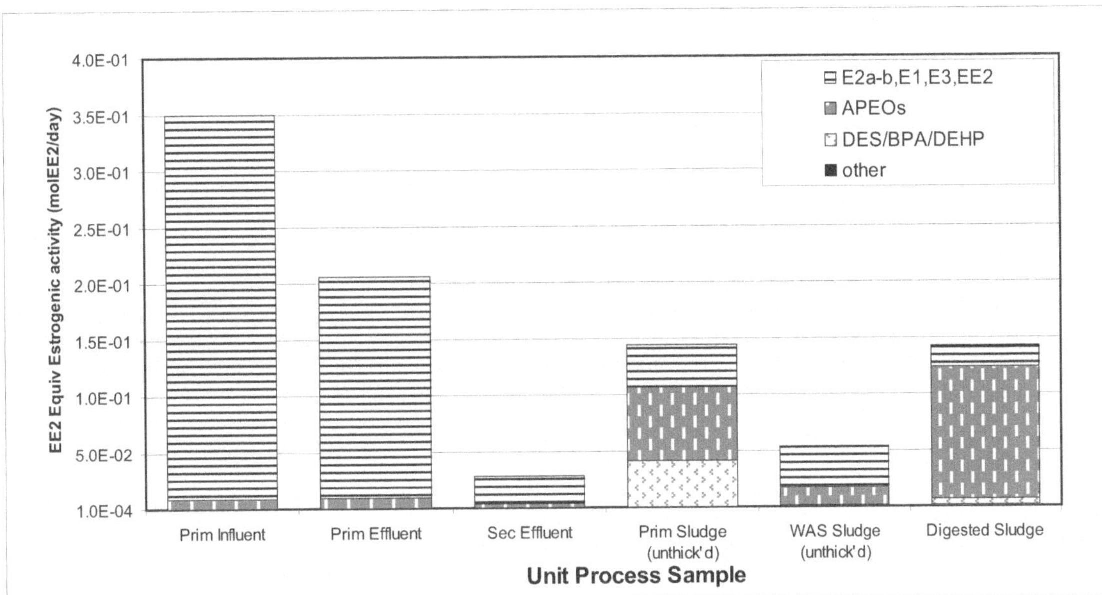

Figure 3-32. Average Daily Estrogenicity Mass Fluxes (mol EE2-equivalents/day) at Plant D. (Based on the Model of Concentration Addition – Calculated Using YES Bioassay-Based EE2 Potency Factors (Table 3-1) – June 2006 Data Only)

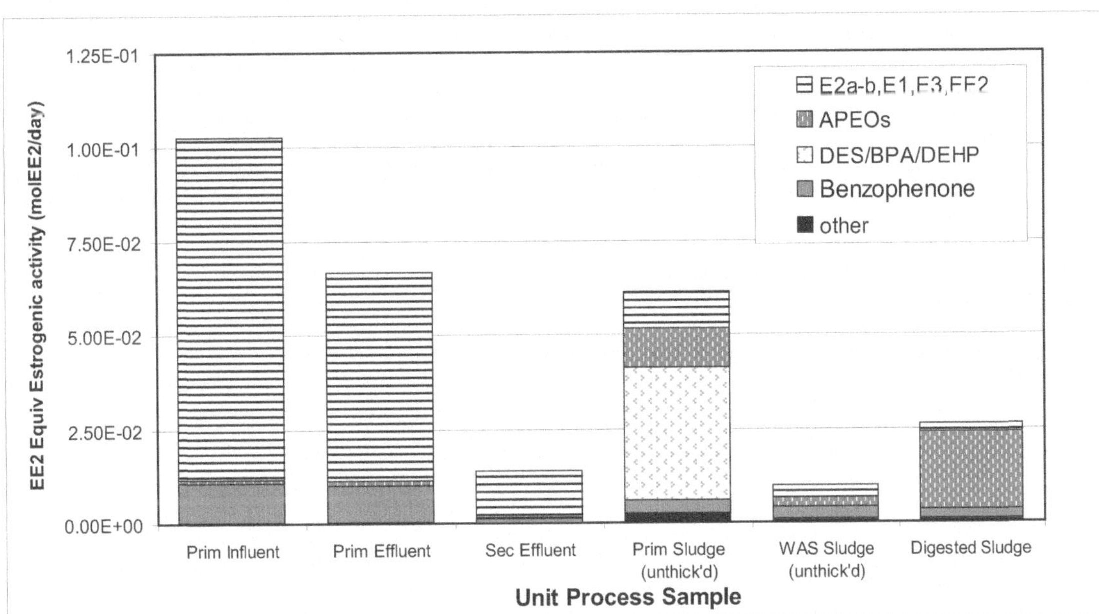

Figure 3-33. Average Daily Estrogenicity Mass Fluxes (mol EE2-equivalents/day) at Plant D. (Based on the Model of Concentration Addition – Calculated Using KBluc Bioassay-Based EE2 Potency Factors (Table 3-1) – June 2006 Data Only)

Daily mass fluxes of estrogenic activity based on the June 2006 YES and KBluc bioassay measurements through unit treatment operations at Plant D are shown in Figures 3-34. The estrogenic activity mass flux reductions across the entire plant were 95% (KBluc) and 76% (YES). During activated sludge treatment, YES-based and KBluc-based mass flux decreases were 51% and 98%, respectively. During thermophilic anaerobic digestion, estrogenic mass flux decreased by 60% (KBluc) and by 62% (YES). The YES and KBluc bioassay results exhibited trends consistent with the Model of Concentration Addition results. Both methods reveal that there was a greater mass flux of estrogenicity discharged from this facility in the solids than was discharged in the secondary effluent, consistent to the finding at Plant B.

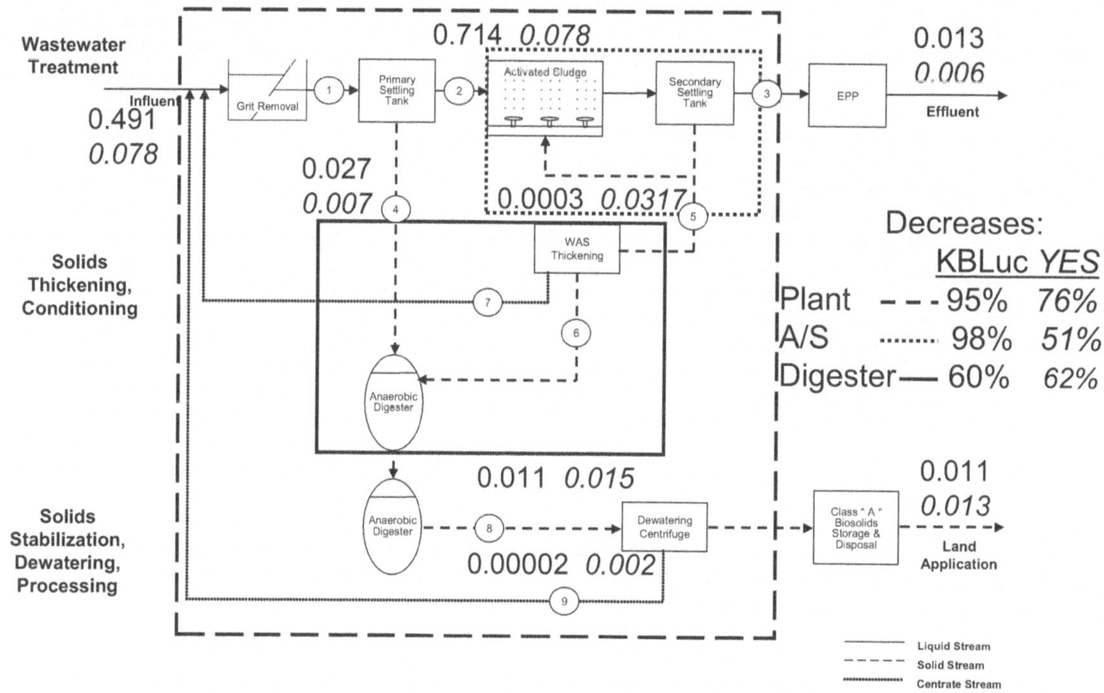

Figure 3-34. Daily Estrogenic Activity Mass Fluxes (mol EE2 equivalents/day) Through Unit Treatment Processes at Plant D. (Based on KBluc and YES Bioassay Measurements – June 2006 Sample Set Only)

3.7 Non-Estrogenic TOrCs

3.7.1 Pharmaceuticals Frequency and Concentration

In addition to hormone and synthetic compounds that exhibit endocrine disrupting activity, 19 non-estrogenic pharmaceuticals were determined in liquid and solid phases from all four plants included in this study. The reason for including the pharmaceuticals was that sources and pathways of introduction are similar to those of EDCs contributed to wastewater, and as biologically active compounds they may be of inherent interest, or as synergists or antagonists of EDCs.

Typical concentrations for any pharmaceutical varied substantially by plant and by unit process, within and between any one matrix (Furlong et al., 2010). From the perspective of chemical analysis, this variability is an inherent aspect of working with wastewater liquid and solid samples, since complete isolation of the variety of pharmaceuticals of interest is not practical or potentially achievable and the range of pharmaceutical chemistries included in this survey predicate against uniformly optimal performance of a single method for all compounds. Coupled with the extremely wide range of concentrations observed within and between liquid phases, caution is necessary in comparing either aggregate or individual results in the Tables in this report or the appendices; to reflect the uncertainty results are reported to at most two significant figures. Additional uncertainty is introduced to the results by the scale of the different processes samples, the geographic spread of the four plants sampled, and the necessity of collecting samples using multiple sample collectors and shipping samples by overnight express. Even with the training provided to sample collectors and the use of consistent sample collection and quality assurance/quality control protocols, the scale of the processes sampled and collection of samples by different staff at each plant, invariably introduces additional error, particularly when determining loads and comparing between plants. However, even with these caveats, distinct trends are observed in the data for individual pharmaceuticals. To better focus these trends, calculation of the mean, median, and standard deviation of concentrations of individual pharmaceuticals, sorted by unit process, are summarized in Table 3-17.

Table 3-17. Summary of Mean Pharmaceutical Concentrations (Solids: ng/g, Liquids: ug/L) of all Four Plants, for Each Unit

Compound		Primary Influent	Primary Effluent	Secondary Effluent	Centrate Recycle Stream	TWAS Centrate	Dewatering Centrate	Thickened Sludge (Combined Primary and Secondary)	Primary Unthickened Sludge	Secondary Unthickened Sludge	Aerobically(1) and Anaerobically Digested Sludge	Acid Phase Digested Sludge	Methane Phase Digested Sludge	Nit / Denit Waste Sludge	Conventional Digested Sludge	Dewatered Sludge	Composted,pelletized and waste activated sludge	Digested Sludge	Centrate Recycle Stream	Thickened Wasted Activated Sludge (TWAS)	TWAS Centrate
	Sample Phase----->	Liquid	Liquid	Liquid	Liquid	Liquid	Liquid	Solid	Solid	Solid	Solid	Solid	Solid	Solid	Solid	Solid	Solid	Solid	Solid	Solid	Solid
1,7-dimethylxanthine	Mean	6.01	6.73	0.18	1.29	0.21															
	Median	5.89	7.86	0.18	1.30	0.21															
	Standard Deviation	4.07	4.06		0.96																
	# of detections/# of total observations	4/4	4/4	1/4	3/12	1/3	--	--	--	--	--	--	--	--	--	--	--	--	--	--	--
Acetaminophen	Mean	61.64	37.74		1.98		0.34												0.03		
	Median	18.00	27.75		0.39		0.34												0.03		
	Standard Deviation	98.56	40.20		2.96		0.08														
	# of detections/# of total observations	4/4	4/4	--	3/12	--	2/4	--	--	--	--	--	--	--	--	--	--	--	1/2	--	--
Albuterol (Salbutamol)	Mean				0.36																
	Median				0.15																
	Standard Deviation				0.39																
	# of detections/# of total observations	--	--	--	3/12	--	--	--	--	--	--	--	--	--	--	--	--	--	--	--	--
Caffeine	Mean	57.73	35.91	0.52	11.42	0.36		259.71			13.89	1,677.96			520.60	55.72	550.37	129.34	0.04		67.74
	Median	32.60	32.10	0.52	1.48	0.24		37.28			13.89	1,677.96			520.60	55.72	550.37	129.34	0.04		67.74
	Standard Deviation	69.11	25.97	0.65	15.49	0.21		467.88													
	# of detections/# of total observations	4/4	4/4	2/4	5/12	3/3	--	4/6	--	--	1/6	1/2	--	--	1/1	1/7	1/5	1/3	1/2	--	1/1
Carbamazepine	Mean	3.81	1.93	1.94	0.22	0.16	0.07	14.47	34.09	42.32	5.33	35,744.42	18.32			34.87	96.30		0.00	0.85	2.05
	Median	0.10	0.14	0.15	0.08	0.14	0.07	3.78	3.50	42.32	3.50	35,744.42	18.32			10.98	6.82		0.00	0.85	2.05
	Standard Deviation	6.49	3.66	3.66	0.31	0.06	0.04	19.10	53.49	57.48	3.51					49.28	159.11		0.00		
	# of detections/# of total observations	3/4	4/4	4/4	11/12	3/3	2/4	3/6	3/5	2/3	3/6	1/2	1/1	--	--	3/7	3/5	--	2/2	1/3	1/1
Cimetidine	Mean																				
	Median																				
	Standard Deviation																				
	# of detections/# of total observations	--	--	--	--	--	--	--	--	--	--	--	--	--	--	--	--	--	--	--	--
Codeine	Mean	0.11	0.10	0.13	0.07	0.45	0.03		236.21	170.46						61.49			0.01		
	Median	0.11	0.10	0.04	0.07	0.45	0.03		236.21	170.46						61.49			0.01		
	Standard Deviation			0.16	0.07		0.02														
	# of detections/# of total observations	1/4	1/4	3/4	3/12	1/3	2/4	--	1/5	1/3	--	--	--	--	--	1/7	--	--	1/2	--	--

Table 3-17. Summary of Mean Pharmaceutical Concentrations (Solids: ng/g, Liquids: ug/L) of all Four Plants, for Each Unit Process (continued).

Compound		Primary Influent	Primary Effluent	Secondary Effluent	Centrate Recycle Stream	TWAS Centrate	Dewatering Centrate	Thickened Sludge (Combined Primary and Secondary)	Primary Unthickened Sludge	Secondary Unthickened Sludge	Aerobically(1) and Anaerobically Digested Sludge	Acid Phase Digested Sludge	Methane Phase Digested Sludge	Nit / Denit Waste Sludge	Conventional Digested Sludge	Dewatered Sludge	Composted,pelletized and waste activated sludge	Digested Sludge	Centrate Recycle Stream	Thickened Wasted Activated Sludge (TWAS)	TWAS Centrate
	Sample Phase----->	Liquid	Liquid	Liquid	Liquid	Liquid	Liquid	Solid	Solid	Solid	Solid	Solid	Solid	Solid	Solid	Solid	Solid	Solid	Solid	Solid	Solid
Cotinine	Mean	0.56	0.54	0.19	0.18	0.26	0.09														
	Median	0.64	0.57	0.19	0.17	0.26	0.09														
	Standard Deviation	0.16	0.14		0.12		0.00														
	# of detections/# of total observations	3/4	3/4	1/4	7/12	1/3	2/4	--	--	--	--	--	--	--	--	--	--	--	--	--	--
Dehydronifedipine	Mean			0.01																	
	Median			0.01																	
	Standard Deviation			0.01																	
	# of detections/# of total observations	--	--	3/4	--	--	--	--	--	--	--	--	--	--	--	--	--	--	--	--	--
Diltiazem	Mean			0.11	0.10	0.21		0.30	2.37	0.40							0.63		0.00		0.90
	Median			0.12	0.06	0.21		0.30	2.37	0.40							0.63		0.00		0.90
	Standard Deviation			0.03	0.09												0.18		0.00		
	# of detections/# of total observations	--	--	3/4	3/12	1/3	--	1/6	1/5	1/3	--	--	--	--	--	--	2/5	--	2/2	--	1/1
Diphenhydramine	Mean	1.58	0.73	0.13	0.07	0.26		130.70	30.38	50.23	16.72	1,170.10		2.55	360.27	94.32	501.97	2.00	0.01	7.23	1.84
	Median	1.58	0.73	0.13	0.06	0.26		24.93	26.33	11.51	17.46	1,170.10		2.55	360.27	29.33	25.06	2.00	0.01	2.84	1.84
	Standard Deviation	2.19		0.02	0.05	0.01		254.16	29.24	67.54	10.46	1,477.58				118.36	955.74	2.57	0.01	9.49	
	# of detections/# of total observations	2/4	1/4	4/4	6/12	2/3	--	5/6	5/5	3/3	4/6	2/2	--	1/1	1/1	7/7	5/5	2/3	2/2	3/3	1/1
Erythromycin	Mean												176.03								
	Median												176.03								
	Standard Deviation																				
	# of detections/# of total observations	--	--	--	--	--	--	--	--	--	--	--	1/1	--	--	--	--	--	--	--	--
Fluoxetine	Mean		0.38		0.00			0.58	4.50	1.63	2.16			32.72		0.82	469.29		0.00	3.67	
	Median		0.38		0.00			0.58	4.50	1.63	2.16			32.72		1.10	5.04		0.00	3.67	
	Standard Deviation							0.43								0.70	806.02			1.48	
	# of detections/# of total observations	--	1/4	--	1/12	--	--	2/6	1/5	1/3	1/6	--	--	1/1	--	3/7	3/5	--	1/2	2/3	--
Miconazole	Mean							92.31	66.78	138.87	25.16	279.51		47.26	445.05	129.17		12.59			
	Median							17.31	38.83	77.43	33.86	279.51		47.26	445.05	48.33		12.59			
	Standard Deviation							181.97	62.00	162.20	16.74					174.31		5.87			
	# of detections/# of total observations	--	--	--	--	--	--	6/6	5/5	3/3	5/6	1/2	--	1/1	1/1	7/7	--	2/3	--	--	--

Table 3-17. Summary of Mean Pharmaceutical Concentrations (Solids: ng/g, Liquids: ug/L) of all Four Plants, for Each Unit Process (continued).

Compound		Primary Influent	Primary Effluent	Secondary Effluent	Centrate Recycle Stream	TWAS Centrate	Dewatering Centrate	Thickened Sludge (Combined Primary and Secondary)	Primary Unthickened Sludge	Secondary Unthickened Sludge	Aerobically(1) and Anaerobically Digested Sludge	Acid Phase Digested Sludge	Methane Phase Digested Sludge	Nit / Denit Waste Sludge	Conventional Digested Sludge	Dewatered Sludge	Composted,pelletized and waste activated sludge	Digested Sludge	Centrate Recycle Stream	Thickened Wasted Activated Sludge (TWAS)	TWAS Centrate
	Sample Phase----->	Liquid	Liquid	Liquid	Liquid	Liquid	Liquid	Solid	Solid	Solid	Solid	Solid	Solid	Solid	Solid	Solid	Solid	Solid	Solid	Solid	Solid
Ranitidine	Mean			0.11	0.10	0.33	0.16														
	Median			0.11	0.10	0.33	0.16														
	Standard Deviation			0.10	0.11																
	# of detections/# of total observations	--	--	2/4	2/12	1/3	1/4	--	--	--	--	--	--	--	--	--	--	--	--	--	--
Sulfamethoxazole	Mean	0.09	0.11	0.07	0.02								51.83								
	Median	0.09	0.11	0.07	0.02								51.83								
	Standard Deviation	0.02	0.03	0.01																	
	# of detections/# of total observations	2/4	2/4	3/4	1/12	--	--	--	--	--	--	--	1/1	--	--	--	--	--	--	--	--
Thiabendazole	Mean							182.56													
	Median							182.56													
	Standard Deviation																				
	# of detections/# of total observations	--	--	--	--	--	--	1/6	--	--	--	--	--	--	--	--	--	--	--	--	--
Trimethoprim	Mean	3.96	0.82	0.07	0.06	0.20											0.58		0.00		1.59
	Median	3.96	0.82	0.07	0.06	0.20											0.58		0.00		1.59
	Standard Deviation			0.02		0.04															
	# of detections/# of total observations	1/4	1/4	4/4	1/12	2/3	--	--	--	--	--	--	--	--	--	--	1/5	--	1/2	--	1/1
Warfarin	Mean						0.10	0.00													
	Median						0.10	0.00													
	Standard Deviation																				
	# of detections/# of total observations	--	--	--	--	--	1/4	1/6	--	--	--	--	--	--	--	--	--	--	--	--	--
Carbamazepine-d10	Mean	32.23	33.65	32.00	10.53	108.40	35.90	0.05	14.07	0.32	13.57	0.36	0.36	0.01		0.16	15.63	16.76	0.00	19.53	31.71
	Median	28.90	33.30	34.75	7.40	90.10	17.85	0.02	1.05	0.11	0.06	0.36	0.36	0.01		0.06	8.23	16.76	0.00	21.73	31.71
	Standard Deviation	21.74	21.75	11.24	8.79	47.85	45.78	0.07	20.25	0.46	30.12					0.28	21.68	9.12	0.00	18.45	
	# of detections/# of total observations	4/4	4/4	4/4	12/12	3/3	4/4	4/6	5/5	3/3	5/6	1/2	1/1	1/1	--	5/7	4/5	2/3	2/2	3/3	1/1
Ethyl Nicotinate-d4 (surrogate)	Mean	44.50	44.23	66.95	29.31	165.13	66.63	0.00	15.31	0.13	19.99	0.25	0.38			0.31	15.09	30.82	0.00	21.48	44.12
	Median	44.40	41.00	60.95	23.65	95.20	63.90	0.00	21.68	0.13	0.10	0.25	0.38			0.31	15.09	29.92	0.00	21.48	44.12
	Standard Deviation	16.12	20.61	23.59	27.74	142.51	37.38		12.99		34.54					0.44		30.88		6.46	
	# of detections/# of total observations	4/4	4/4	4/4	12/12	3/3	4/4	1/6	3/5	1/3	3/6	1/2	1/1	--	--	2/7	1/5	3/3	1/2	2/3	1/1

First, at least one pharmaceutical was detected in all samples, although some pharmaceuticals were detected infrequently. Overall diphenhydramine, carbamazepine, miconazole, and caffeine were the most frequently detected compounds in all samples (combined liquids and solids) at detection frequencies of 73, 65, 40, and 39%, respectively (Table 3-18). A number of compounds were detected in less than 10% of samples, with some pharmaceuticals detected in only one sample or in a few samples from one unit process. For example, albuterol was detected only in centrate recycle liquids and then in 3 of 12 of those samples analyzed over the course of the study. Erythromycin and thiabendazole were detected once in one sample of methane-phase digested sludge or thickened sludge, respectively. The only pharmaceutical measured that was not detected in any samples in this study was cimetidine.

As can be seen in Table 3-18, the frequency of detection for other pharmaceuticals varied between the frequent observations made for diphenhydramine, carbamazepine, miconazole, and caffeine, and the very infrequent detections of albuterol, erythromycin, and thiabendazole, but there appears to be significant differences in the distribution of pharmaceuticals found in liquids versus those found in solids (Table 3-17).

Table 3-18. Overall Frequency of Occurrence of All Pharmaceuticals in All Media from All Unit Processes in Plants A-D.

Compound	Frequency of detection
Diphenhydramine	72.7%
Carbamazepine	64.9%
Miconazole	40.3%
Caffeine	39.0%
Cotinine	22.1%
Fluoxetine	22.1%
Codeine	19.5%
Diltiazem	19.5%
Acetaminophen	18.2%
1,7-dimethylxanthine	16.9%
Trimethoprim	15.6%
Sulfamethoxazole	11.7%
Ranitidine	7.8%
Albuterol (Salbutamol)	3.9%
Dehydronifedipine	3.9%
Warfarin	2.6%
Erythromycin	1.3%
Thiabendazole	1.3%
Cimetidine	0.0%
Carbamazepine-d10	88.3%
Ethyl Nicotinate-d4 (surrogate)	66.2%

3.7.1.1 Liquid Samples

In liquids, pharmaceuticals are more frequently detected and more individual pharmaceuticals are detected in any of the liquid-phase unit processes when compared to solid phase processes. For example, nine of 19 pharmaceuticals were detected in at least one of four primary influent liquid samples, with similar trends observed in primary and secondary effluent (10 of 19 pharmaceuticals and 11 of 19 pharmaceuticals, detected in at least one of four samples, respectively).

Interestingly, three pharmaceuticals, dehydronifedipine, diltiazem, and ranitidine, were detected in multiple (two or more of four) samples of secondary effluent at median concentrations of 0.014, 0.12, and 0.11 µg/L, respectively, when these compounds went undetected in either primary influent or effluent. As also can be seen in Table 3-16 median concentrations of many pharmaceuticals present in both primary influent and effluent remained approximately the same, although concentrations of some compounds, including diphenhydramine, sulfamethoxazole, and trimethoprim, appear to be reduced by 50% or more.

In comparison, reduction of liquid concentrations of pharmaceuticals is most pronounced after secondary treatment. Concentrations of seven of nine pharmaceuticals detected in primary influent, primary effluent and secondary effluent decrease by 60% or more between primary and secondary treatment processes (Table 3-17 and Figure 3-35). One compound, acetaminophen, was reduced to undetectable levels in secondary treatment.

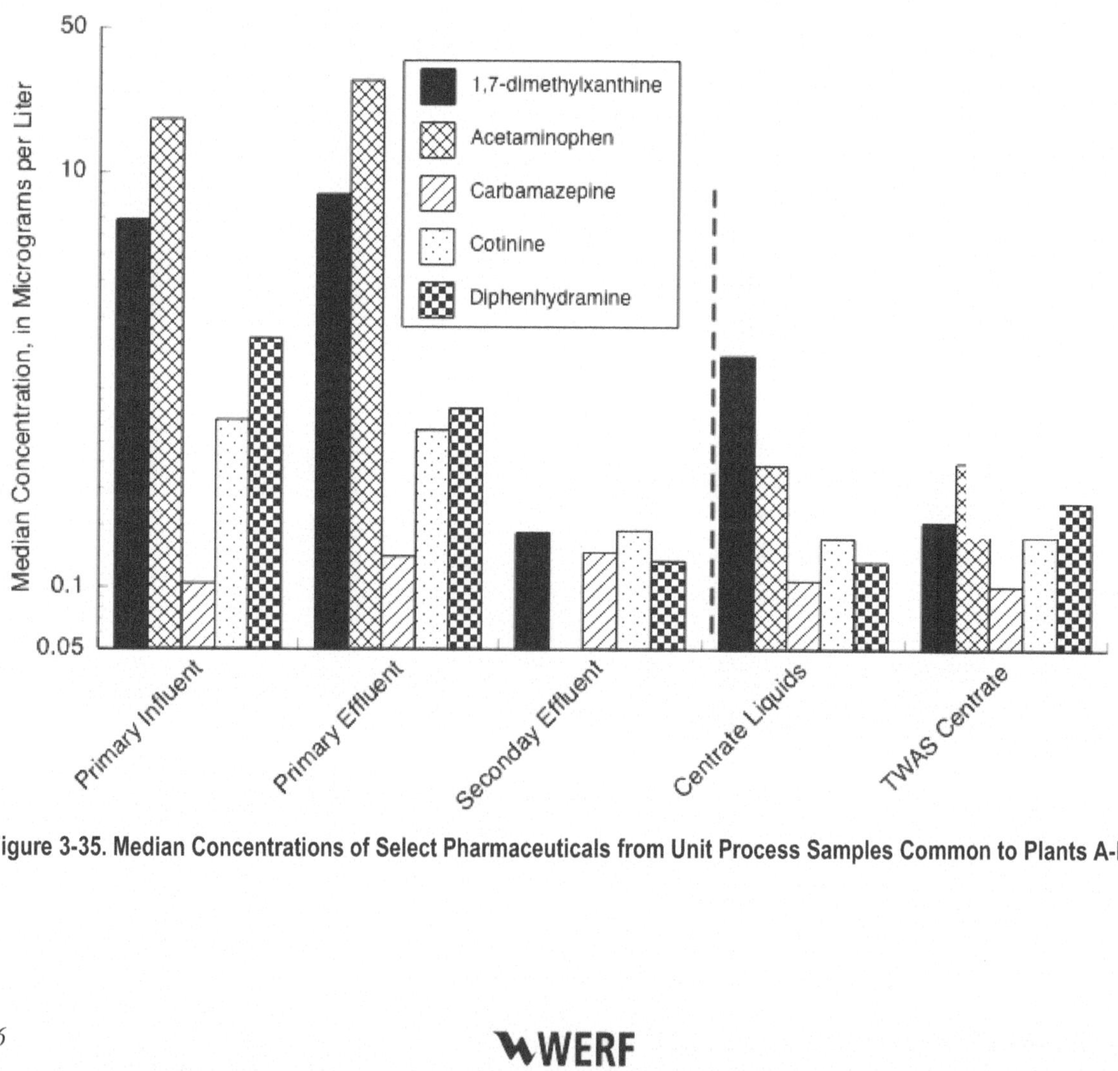

Figure 3-35. Median Concentrations of Select Pharmaceuticals from Unit Process Samples Common to Plants A-D.

It is important to note an important exception to this reduction. Carbamazepine concentrations are essentially constant between primary influent and secondary effluent, with similar concentrations (within a factor of 2) in the internal liquid cycling processes, such as the centrate streams that are return flows to secondary treatment.

An important analytical aspect that may influence the magnitude of these observed changes is that as the liquid flow moves from primary influent to secondary effluent, reduction of organic matter, particularly during secondary treatment, removes a component that may cause interferences and affect method performance. This is illustrated by the improvement of median ethyl nicotinoate-d4 (first method surrogate) recoveries through primary influent → primary effluent → secondary effluent, which were 44, 41, and 61%, respectively. A more modest improvement of carbamazepine-d10 (second method surrogate) recoveries (29, 33, and 35, respectively) also was observed, although this increase in recoveries is not significant. Nevertheless, substantial reduction and removal of many pharmaceuticals from the liquid stage is apparent, even with the observed changes in method performance between liquid treatment stages suggested by surrogate recoveries. As noted above, acetaminophen was completely removed by secondary treatment, and 1,7, dimethylxanthine (a caffeine metabolite), caffeine, and trimethoprim were reduced by more than 90%, suggesting effective removal for a number of pharmaceuticals.

The appearance of low concentrations of dehydronifedipine, diltiazem, and ranitidine in secondary effluent but not in primary influent or effluent could result from at least two effects: 1) better detectability in secondary effluent samples that have lower total extractable organic matter and thus are less prone to interferences that reduce recovery or impede detection, and 2) a true increase in these compounds, which may result from breakdown (during secondary biological treatment) of labile conjugated metabolites of these compounds that are excreted in human waste and thus were not detectable or available for extraction in the primary influent or effluent samples. It may be that both of these effects are in play in these treatment plants, but until analytical methods capable of detecting a range of commonly excreted conjugated forms of many pharmaceuticals is available, the relative importance of each of these two effects cannot be separated.

3.7.1.2 Solid Samples

In contrast to the liquid samples, fewer pharmaceuticals were routinely detected in solids samples, regardless of which step in treatment process was sampled. However, on a mass basis, compound concentrations were typically much higher in solid samples than in corresponding liquid samples (Table 3-17). For example in primary influent, median concentrations of the nine pharmaceuticals detected ranged between 0.09 and 33 μg/L (ppb), whereas the concentrations of the six pharmaceuticals detected in final sludge products such as composted or pelletized sludge, ranged between 6.8 and 550 μg/kg (ppb); this range of solids concentrations is similar to that observed by other investigators (Kinney et al., 2006a, Radjenovic et al., 2009).

Caffeine, carbamazepine, diphenhydramine, and miconazole were frequently detected in both solids and liquids; miconazole, the fourth most frequently detected compound, was detected in solids only. Fluoxetine was frequently detected in solids, with fewer detections of codeine, diltiazem, and trimethoprim (Table 3-17). Sulfamethoxazole, typically combined with trimethoprim when prescribed was detected in one solids sample, at 51 μg/kg, higher than any of the solids detections of trimethoprim, and a reverse of the observed concentrations in liquid samples, where concentrations of trimethoprim typically were higher than sulfamethoxazole. Acetaminophen was detected once at low (0.03 μg/L) concentration in a centrate recycle stream solids sample.

The observation of a smaller range of pharmaceuticals in solids, but at typically higher concentrations may be explained by the higher concentration of organic matter present in solids samples may result in higher concentrations of compounds that are likely to sorb to organic matter. Increasing values of the log octanol-water partitioning coefficient are reflective of a greater likelihood to partition to organic-carbon rich solids, and the Octanol-Water Partition Coefficients (log K_{ow}) for carbamazepine, diphenhydramine, fluoxetine and miconazole have been reported for these compounds as 2.45, 3.27, 4.05, and 6.25 (Kinney et al., 2006b). The log K_{ow} of caffeine (-0.07) suggests that caffeine should not be associated with organic-carbon rich sludges, but the high concentrations observed for some samples, ranging between 0.04 and 1,700 μg/kg, but more typically ranging between 14 and 550 μg/kg, lends evidence to the hypothesis that other factors than partitioning to organic matter controls the distribution of pharmaceuticals between liquid and solid phases.

Pharmaceutical concentrations appear to increase as sludge is processed through treatment. This can be seen in the concentrations of carbamazepine, diphenhydramine, and miconazole as treatment progresses, shown in Figure 3-36. The increases are suggestive that as liquid and solid phases move through treatment, there may be transfer of pharmaceuticals from liquid to solid phase, particularly those with larger log K_{ow} values. While this appears as a net removal from the liquid phase, it poses potential challenges for solids processing, treatment, and disposal.

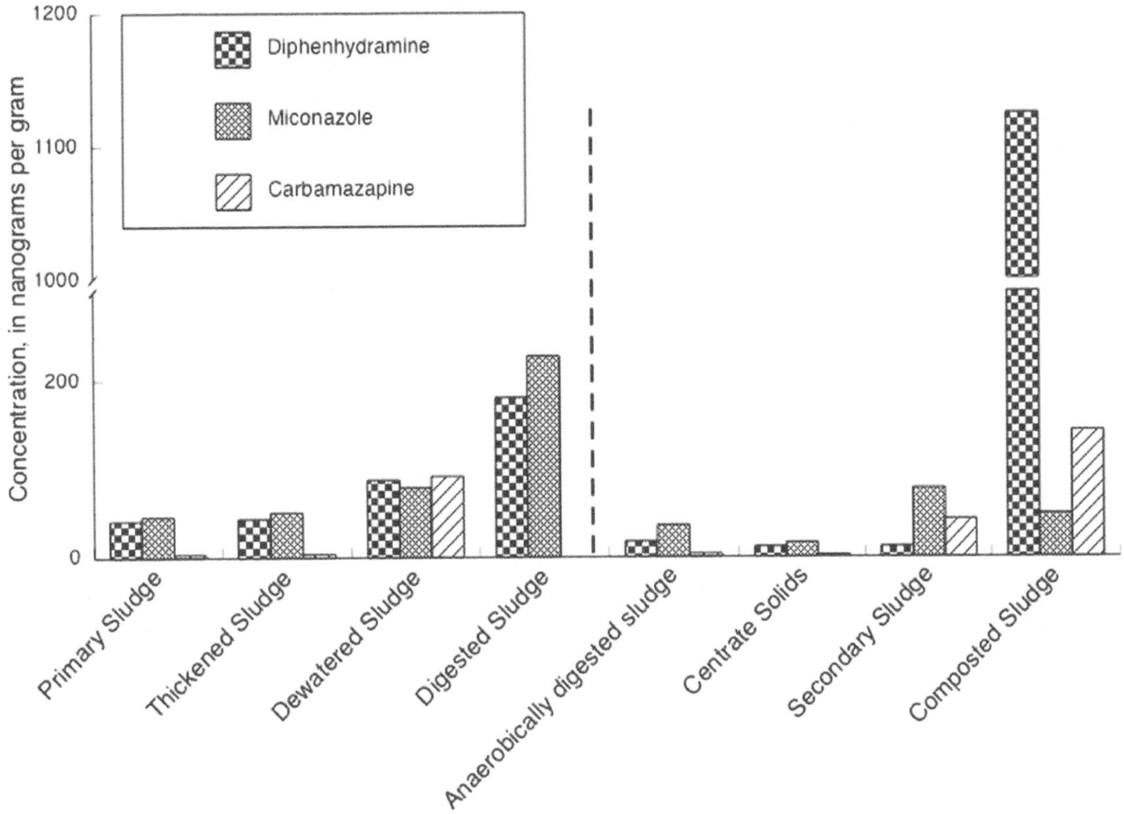

Figure 3-36. Median Concentrations of Pharmaceuticals in Solids from Plants A-D.

When compared to the other liquid and solid samples in this study, the liquid and solid centrate recycle stream samples were unique, both analytically and scientifically. Observation and processing of these samples indicated that they were a difficult-to-separate mixture of very fine flocculant solids and a liquid phase. Upon receipt at the laboratory, the field-filtered samples frequently required additional treatment to separate the solid and liquid phases, either additional filtration or centrifugation. Because of sample size limitations, only two solids samples were analyzed. The results from the liquid centrate recycle stream samples suggest that this return flow liquid contained a wide array of pharmaceuticals (13 of 19 pharmaceuticals detected in at least one of 12 samples). The concentrations observed fell in between concentrations typical of the primary and secondary effluent results (Table 3-17).

Eight of the 19 pharmaceuticals determined were at detectable values in at least one of the two solid centrate recycle stream samples; however, unlike other solids samples, the observed concentrations were relatively low, ranging between 0.005 and 0.04 µg/kg. These low concentrations may result from poor method performance as indicated by surrogate recoveries for these samples, which may reflect the challenges associated with handling and analyzing these samples. However, even with suboptimal method performance, the results suggest that the steps used to dewater and consolidate solids prior to disposal at Plants A and D (sources of the centrate recycle stream samples), such as flocculation and centrifuging, may substantially alter the distribution of pharmaceuticals between liquid and solids phases, and that return flow of this stream to the earlier steps in the overall treatment train is appropriate particularly for the liquid phase.

3.7.2 Instantaneous Loads of Pharmaceuticals from Plants B and D

Pharmaceutical concentrations are indicative of processes occurring during wastewater treatment; however to accurately understand whether observed changes in concentration reflect true removal of a pharmaceutical or transfer between solid and liquid phases, instantaneous loads were determined based on solid and liquid loads calculated elsewhere in this report. Review of the pharmaceutical data in this study indicated that sufficient data to compare liquid and solid phase loadings was available for Plant B in July 2006 and January 2007 and for Plant D for June and December 2006. Table 3-19 contains the instantaneous loads, in grams per day, for these two plants and two sampling events for 5 pharmaceuticals, acetaminophen, caffeine, carbamazepine, diphenhydramine, and miconazole. Note that miconazole could not be determined using the analytical method applied to liquid samples, so comparison between liquid and solid loadings cannot be made. Also note that a final pelletized sludge sample was not available for analysis of pharmaceuticals in June 2006 in plant B, which also may limit interpretations of loadings during that event.

A number of observations can be made with the available pharmaceuticals loading data (Table 3-19). For acetaminophen in Plants B and D during both events, remineralization of acetaminophen after secondary treatment is very efficient, typically 99% or greater of the primary influent or effluent load. As noted in an earlier section, no acetaminophen was detected in solid samples, so solids loadings are effectively zero. Caffeine is similarly well remineralized to a greater 99%, although small but detectable loads are present in solids, in some cases comparable to the loads present in secondary effluent. In plant D, remineralization was greater during the December 2006 sampling event than in June 2006, even though the loading of caffeine in December 2006 primary effluents was a factor of two greater.

Carbamazepine loadings for both plants suggest that relatively little carbamazepine is remineralized in liquid phase during treatment. No specific pattern of reduction could be detected between primary influent, primary effluent and secondary effluent, with the highest loadings of carbamazepine consistently detected in secondary effluent in Plant D and with higher loadings in the primary and secondary effluents in Plant B in summery and higher loadings in primary influent in winter. Plant B also was distinguished with a much higher (by a factor of 20 to 100) loadings of carbamazepine in January 2007 compared to July 2006. Loadings of carbamazepine in solids were a few% or less of the loadings in liquids, indicating that the bulk of carbamazepine leaves wastewater treatment untransformed and almost completely in the liquid phase. This suggests that under current treatment approaches, treated effluent discharge could be a source for the common observed detection of carbamazepine in surface waters globally (Glassmeyer et al., 2008), and indicates the need for further study to develop efficacious means of remineralizing this compound during treatment.

Diphenhydramine also was detected in liquid and solid phases; however, the distribution of instantaneous loads, particularly in plant D, suggests production of diphenhydramine during secondary treatment, with the distribution of loads between is less clearcut. In three of the four sampling events (July 2006 in Plant B, June and December 2006 in Plant D), loadings in the liquid phase are typically between 0 and 180 grams per day. In July in Plant B, diphenhydramine was undetectable in primary influent and effluent, and rose to 74 grams per day in secondary effluent, while in June and December 2006 in Plant D primary influent and effluent loadings range from between 30 to 60 grams per day, increasing to 120 or 180 grams per day in secondary treatment. Most of the diphenhydramine loading in both plants is in the liquid phase although diphenhydramine loads are measurable in most of the intermediate and final sludge processing steps. As noted earlier, the method available for analyzing liquid samples did not provide miconazole results, but they are included to show the potential importance of this compound as a solids tracer. Instantaneous loads ranging between a few tenths to 80 grams of miconazole per day were observed. The log K_{ow} of 6.05 reported for this pharmaceutical suggests that it will likely be found primarily in solids, but additional analytical methods development is necessary to determine if detectable amounts of this compound also can be found in liquids.

The January 2007 sampling event for Plant B bears particular note in terms of the very high liquid phase loadings observed in primary influent and effluent samples for acetaminophen, caffeine, carbamazepine, and diphenhydramine. These loadings were one to three orders of magnitude greater than in the July 06 sampling. It is noteworthy however, that the trends in loadings in primary influent, primary effluent, and secondary effluent were the similar for acetaminophen, caffeine and carbamazepine during both samplings. Only the pattern of loadings for diphenhydramine differed. The observed increases in Plant B in January 2007 may reflect changes in plant operation, seasonal changes in compound use, or other factors that have yet to be identified. In contrast the temporal performance of Plant D is relatively constant in the magnitude of loadings at each process step and consistent in the compound specific trends observed.

The loadings results suggest that while substantial removal of some pharmaceuticals, such as acetaminophen and caffeine are observed, the loadings of some compounds such as miconazole, warrant further investigation. The loading trends observed for carbamazepine and diphenhydramine indicate that these compounds persist through treatment and that process improvement or addition of other treatment steps may be necessary to reduce loads and concentrations of these and similarly recalcitrant pharmaceuticals to substantially lower levels. The overall behavior of pharmaceuticals through the differing treatment steps employed by the plants sampled in this study is similar to that observed by others, and suggests that further steps to reduce the loads of recalcitrant pharmaceuticals in liquid phase processes would be most effective in reducing the total loadings of pharmaceuticals exiting treatment plants.

Table 3-19. Instantaneous Loads for Select Pharmaceuticals from Plants B and D.

Sample Location	Matrix		Acetaminophen		Caffeine		Carbamazepine		Diphenhydramine		Miconazole
Plant B											
July 2006											
Primary Influent	Liquid	E	920	E	3,900	E	61	<	13	*	
Primary Effluent	Liquid	E	1,000	E	5,500	E	120	<	13	*	
Secondary Effluent	Liquid	E	14	<	8.6	E	110	E	74	*	
Secondary Unthickened Sludge	Solid	ND		ND		ND			0.71		5
Anaerobically Digested Sludge	Solid	ND		ND		ND		ND		ND	
Acid Phase Digested Sludge	Solid	ND		0	26	ND			34	ND	
Centrate Recycle Stream (Pelletech) (liquid)	Liquid	E	5.7	E	22	E	1.5	<	0.34	*	
January 2007											
Primary Influent	Liquid		120,000		93,000		6,600	E	1,800	*	
Primary Effluent	Liquid		54,000		40,000		4,200	E	410	*	
Secondary Effluent	Liquid	E	14	<	8.6		4,300		81	*	
Primary Unthickened Sludge	Solid	ND		ND			8.7		5.1		4.8
Secondary Unthickened Sludge	Solid	ND		ND			5.3		8.2		21
Anerobically Digested Sludge	Solid	ND		ND			0.32		0.34		1.3
Dewatered Sludge (Pelletech)	Solid	ND		ND			7.1		17		33
Centrate Recycle Stream (Pelletech) (liquid)	Liquid	E	1.4	<	1.2		16		0.41	*	
Plant D											
June 2006											
Primary Influent	Liquid	E	19,000	E	28,000	E	33	E	43	*	
Primary Effluent	Liquid	E	22,000	E	30,000	E	34	<	30	*	
Secondary Effluent	Liquid	<	30	E	1,200	E	39	E	120	*	
Primary Sludge (Unthickened)	Solid	ND		ND			1.3		29		80
Waste Activated Sludge (Unthickened WAS)	Solid	ND		ND		ND			3		6.6
Thickened Wasted Activated Sludge (TWAS)	Solid	ND		ND			0.11		2.4		5.7
TWAS Centrate	Liquid	<	0.8		20		3.3		8.3	*	
Digested Sludge	Solid	ND		ND		ND		ND		ND	
Dewatering Centrate	Liquid		2.8	<	0.15		0.37	<	0.23	*	
December 2006											
Primary Influent	Liquid	E	26,000	E	53,000	<	68	<	38	*	
Primary Effluent	Liquid	E	50,000	E	53,000	E	100	<	60	*	
Secondary Effluent	Liquid	<	30	E	71	E	130	E	180	*	
Primary Sludge (Unthickened)	Solid	ND		ND		ND			1.3		14
Waste Activated Sludge (Unthickened WAS)	Solid	ND		ND		ND			0.098		2.7
Thickened Wasted Activated Sludge (TWAS)	Solid	ND		ND		ND			0.34		4.4
TWAS Centrate	Liquid	<	2.7	E	6.4		4.1	<	2.6	*	
TWAS Centrate	Solid	ND			3.6		0.11		0.097		0.19
Digested Sludge	Solid	ND		ND		ND			0.97		4.3
Dewatering Centrate	Liquid	<	0.84	<	0.52	<	0.63	<	0.8	*	

Notes: E = Estimated, ND = Not Detected, NA = Not Applicable, D-R = , U-D = Result not reported because of sample interferences, Q-D = Result not reported because of result did not meet quality criteria, < = Denotes that the analyte was not detected; the associated parameter value is generally the reporting limit, * = Sample was not analyzed for this compound

CHAPTER 4.0

SUMMARY AND CONCLUSIONS

This was the first research on TOrCs in sludge and biosolids supported by WERF. The primary objective of this study was to provide key baseline information concerning the estrogenicity (measured with *in vitro* bioassays) and concentrations of individual estrogenic TOrCs (measured using GC/MS/MS and LC/MS methods) through common wastewater treatment processes used to condition, thicken, stabilize, and dewater sludge.

Four full-scale WWTPs were sampled between two and four times during a one year period, Plants A, B, C, and D. Biosolids stabilization processes were of particular interest for this project: Plant A uses aerobic digestion; Plant B uses mesophilic anaerobic digestion with both conventional and egg-shaped digesters as well as a two-stage acid-phase digestion process, all operating in parallel trains; Plant C uses lime addition; and Plant D uses thermophilic anaerobic digestion.

Over the course of the study, 15 sample trips were conducted and a total of 90 samples were collected from the study plants. A suite of 100 TOrCs, including steroidal hormones, pharmaceuticals, and AWIs was analyzed. In addition, total estrogenic activity was measured using the YES bioassay, and for selected samples at one plant, the KBluc bioassay.

Due to the extensive amount of data generated in this study, analytical results for both chemical and bioassay analysis were compiled and published as a separate USGS Data Report (Furlong et al., 2010) that is available on the USGS website (http://pubs.er.usgs.gov/).

The instantaneous loads, in g/day, of TOrCs and estrogenic activity were calculated for each sample point based on flows and solids loadings data provided by the study plants. The instantaneous loads of hormones, alkylphenolic compounds and bioassays were presented in this report. The estrogenicity of samples, in EE2-Eqs, was considered from two perspectives: based on the Model of Concentration Addition using the concentrations of individual compounds as measured by USGS and based on the total estrogenicity exhibited by the YES bioassay.

A high degree of variability between sampling dates and within and among the plants was observed, complicating the ability to make conclusive interpretations. This variability can be attributed to error associated with: plant data; closing the mass balance of flow and solids across the unit operations and the interconnected network of flows, sidestreams and recycle streams at the study plants; and chemical and biological analysis of complex solids samples. Furthermore, concentrations of TOrCs were often at or near analytical limits of detection which affected interpretation of increases and decreases in loads. The research team committed to making the best possible interpretations based on the sometimes ambiguous results obtained for each plant, the highlights of which are summarized herein.

Based on the Model of Concentration Addition, nearly all of the estrogenicity derived from compounds that were measured in this study, in all plants and all dates stems from the presence of the 16 compounds listed in Table 4-1 (along with YES potency factors relative to

EE2). The list includes natural and synthetic hormones, alkylphenolic compounds and various other estrogenic compounds.

Table 4-1. Primary Contributors to Estrogenicity (Potency Relative to YES Bioassay).

Compound	Abbreviation	Log K_{OW}	Mol. Wt. (g/mol)	EE2 Equivalents (mol_{EE2}/mol)
17-alpha-ethinyl-estradiol	EE2	4.15	296.39	1.000000
17-alpha-estradiol	E2a	3.67	272.37	0.840000
17-beta-estradiol	E2b	3.94	272.37	0.840000
Estrone	E1	3.43	270.35	0.319000
Estriol	E3	2.81	288.37	0.002000
Diethylstilbestrol	DES	5.07	268.34	0.924000
4-n-octylphenol	4nOP	5.50	206.33	0.000360
4-tert-octylphenol	4tOP	5.28	206.33	0.000360
4-octylphenol monoethoxylates	OP1EO		250.36	0.000010
4-octylphenol diethoxylates	OP2EO		294.42	0.000010
4-nonylphenol	NP	5.92	220.34	0.000010
4-nonylphenol monoethoxylates	NP1EO	4.17	264.39	0.000001
4-nonylphenol diethoxylates	NP2EO	4.21	290.43	0.000001
Bisphenol A	BPA	3.64	228.28	0.000563
Benzophenone	benzoph	3.15	182.22	0.000168
Diethylhexyl phthalate	DEHP	8.39	390.56	0.000021

It is important to note that the absence of DES would be expected, however, it was the largest component of the estrogenic signal in digested solids for one of the plants (A). Due to improvements in GC/MS/MS analysis of DES over the course of the study, and the lack of a likely major source of DES, the confidence in this conclusion and detections at the other study plants is less than for the other hormones and estrogenic TOrCs.

With regard to steroid hormones, it is well documented that estrone (E1) is a metabolite of estradiol (E2) that is readily formed during aerobic treatment processes (e.g., Ternes et al., 1999). Given the high levels of E1 observed in some biosolids samples, it is unlikely that E1 and E2 present in plant influent could account for all of the mass. Therefore, the high levels of estriol (E3) in the influent are of particular interest because E3 represents another source of material that could potentially be transformed to E1. Although E1 is less estrogenic than E2, it is significantly more estrogenic than E3 in the YES bioassay as well as *in vivo* for most fish species that have been tested (Vajda et al., 2008), so if E3 metabolism is a significant contributor to in-plant E1 production it is possible that in some cases biosolids treatment could increase total estrogenicity of a waste stream while decreasing the total concentration of steroidal estrogens. Also, transformation during treatment is generally more complete for androgens than for estrogens, likely due to their lack of an aromatic ring which may be more resistant to transformation.

The distribution of TOrCs in untreated streams is very different from secondary effluent streams. Focusing on the steroids: cis-androsterone, dihydrotestosterone, testosterone, 11-ketotestosterone, and progesterone are rarely observed in treated effluents or surface waters, while androstenedione and estriol are often present in effluents at low levels (i.e., < 10 ng/L). All of these compounds are present in primary influents at levels in the 100s of ng/L and are removed from the aqueous stream with great efficiency. Testosterone, dihydrotestosterone, 11-ketotestosterone, and estriol are not present at particularly high levels in biosolids samples, assessment of instantaneous loads of these compounds indicates they are mostly transformed rather than partitioned into the solid phase. Conversely, cis-androsterone, androstenedione, estrone, and progesterone are present at relatively high concentrations in biosolids, indicating persistence through treatment or potential formation within the plant. For instance, at Plant D in March 2006, the load of estrone in primary sludge and TWAS were 9.1 and 6.2 g/day, respectively, compared with 130 g/day post-digestion.

The alkylphenols contributed strongly to estrogenicity as well as the steroids, particularly in the solid phase. Although they are far less potent than the steroids, they are more hydrophobic, partition more readily into the solid phase, and are generated from an unquantified fraction of longer chain APEOs. Most available literature focuses on the fate of estrogenic steroids in aqueous treatment streams. In these studies, the bulk of estrogenicity is attributed to relatively few compounds, primarily the steroids 17β-estradiol, ethinyl estradiol, estrone, and to a lesser extent, estriol. Alkylphenols, alkylphenol ethoxylates, bisphenol A, and other non-steroidal estrogenic compounds are typically present in treated effluents at μg/L levels (compared to ng/L for hormones). However, their relative activity is such that outside of a few well-documented special cases (e.g., Sheahan et al. 2002) their contribution to total estrogenicity of effluents is considered minimal. This is not the case with solid samples analyzed in this study, especially with respect to the alkylphenols, which are more hydrophobic than the steroidal estrogens.

For all plants in this study, loads of hormones were substantially less than the loads of alkylphenolic compounds in the solids streams for most analytes. For example, at Plant A, the instantaneous loads of alkylphenolic compounds in thickened sludge ranged from 32 to 1,800 g/day whereas hormones loads ranged from non-detect to 0.25 g/day (excluding coprastonal and cholesterol). Therefore, in assessing the efficacy of reduction in estrogenicity, it is important to focus on these constituents in addition to the steroids. In this context, it is important to note that certain digestion processes are effective at removal of APEOs.

While the solids process trains varied across the study plants, one commonality is that each uses activated sludge for secondary treatment of the liquid stream. Activated sludge treatment of the primary effluent significantly decreased estrogenicity. More than 90% of most estrogenic TOrCs were removed from the liquid phase during activated sludge treatment and most of the total estrogenicity in liquids was due to steroidal hormones. Significant decreases in concentrations of TOrCs through activated sludge treatment are well documented in the literature.

The results of this project, as well as published studies by other researchers, suggest that the effectiveness of biosolids digestion in reducing estrogenicity and other TOrC concentrations is highly variable. For Plant A, which uses aerobic digestion, the load of most estrogenic TOrCs decreased through the digestion process. Loads of most hormones were non-detect or very low. There were substantially higher loads of alkylphenolic compounds in the thickened sludge (32 to 1,800 g/day), which were reduced following digestion but not to non-detect levels (8.7 to 690 g/day). The loads of TOrCs and their reduction through digestion corresponded with the 18% reduction observed using the YES bioassay. Compared to aerobic digestion, mesophilic (Plant B) and thermophilic (Plant D) anaerobic digestion caused the estrogenic load, as measured by the YES bioassay, to increase. This was likely a consequence of an increased contribution by alkylphenols, particularly nonylphenol (NP), which is more estrogenically potent than its ethoxylated precursors. NP is largely removed during aerobic processes. The magnitude of the estrogenicity increase during anaerobic digestion seems to correlate with digestion temperature and/or the amount of alkylphenol degradation that may have occurred in the collection system piping prior to entering the WWTP.

Several trends were observed at Plant B. First, the loads of hormones generally decreased from plant influent to finished biosolids and effluent. For instance in July, loads of E1, E2, and E3 all decreased substantially from the primary effluent stream (57, 18, and 300 g/day, respectively) to both secondary effluent (<0.46 for each) and dewatered sludge (5.1, ND, and 1.4 g/day, respectively). Conversely, loads of alkylphenolic compounds post-digestion were more variable and likely due to the degradation of longer chain NPEOs and formation of shorter chain NPEOs. Overall, the YES bioassay measurements show similar trends as predicted with the Model of Concentration Addition. Similar to TOrC loads, there is a large decrease in estrogenicity during secondary treatment and the total estrogenicity of the solids increased after mesophilic anaerobic digestion (attributed to APEOs); however this is less than the load coming into the plant. The mass flux of estrogenic activity increased from 0.486 to 0.638 mmol EE2-eqs/day, an increase of 31% during anaerobic digestion. The chemical and bioassay measurements both reveal that there is a greater amount of estrogenicity discharged from this facility in the solids than in the secondary effluent. Finally, on the sample date both the composted and pelletized products were analyzed, low loads of total estrogenicity of 0.077 and 0.0073 g/day of EE2-eqs, respectively, were calculated.

At Plant D, instantaneous load results for both hormones and alkylphenolic compounds were variable following digestion over the sample dates. The Model of Concentration Addition predicted that estrogenicity in primary influent was dominated by hormones whereas estrogenicity in treated biosolids was dominated by APEOs. The YES bioassay results were most comprehensive in June 2006 and December 2005 and showed a reduction in estrogenic load through the plant (both biosolids and final effluent) despite an increase in load following digestion. The chemical and bioassay measurements both reveal that there was a greater load of estrogenicity discharged from this facility in the solids than was discharged in the secondary effluent, consistent with the finding at Plant B.

The lime stabilization process used at Plant C removed more than 90% of alkylphenols during July, although it was less effective during the winter. Over both sample dates, loads of the majority of steroids decreased post-lime from loads in the dewatered sludge. In contrast to the decrease of estrogenic compounds, the total estrogenicity, as measured by the YES bioassay, increased dramatically during lime stabilization in both December 2005 and July 2006 from 2.5 to 9.7 g/day and 6.2 to 26 g/day EE2 Eqs, respectively. Sample frequency was not sufficient to further evaluate the reason but it may be due to conversion of untargeted compounds to more estrogenic products during lime stabilization; or possibly by a contribution of an estrogenic contaminant in the lime itself; or an effect of the dramatic increase in pH.

Concentrations and instantaneous loads of non-estrogenic TOrCs (e.g. pharmaceuticals), a secondary objective of this study, also were measured in order to determine removal of these

TOrCs during treatment. Overall, removal occurs primarily during secondary treatment (activated sludge), with little or no removal during primary treatment. Results also suggest that particle-mediated removal is likely for several compounds, particularly compounds with high log K_{ow} values, although it is likely that the hydrophobic partitioning reflected by log K_{ow}, is not the only processing mediating compound sorption to solids. Other compounds, such as acetominophen and caffeine, are more effectively remineralized, and removal by activated sludge secondary treatment is an effective means of reducing total influent instantaneous loads. Persistent recalcitrant compounds such as carbamazepine, which is inefficiently remineralized and persists in both solid and liquid phases, pose the greatest challenges to removal during treatment.

A major conclusion from this study is that for all plants the load of estrogenic TOrCs leaving the plant in biosolids (and liquids, for the plants where liquid samples were collected) was less than that entering the plant (or as measured in the least treated sample point (e.g. unthickened sludge)). This was the case even for plants where loads of estrogenic TOrCs and/or estrogenicity increased post-stabilization (e.g. anaerobic digestion). Additionally, although the contribution to total estrogenicity by non-steroidal TOrCs (e.g. alkylphenols) varied from plant to plant, the results indicate they can be a major contributor and cannot be ignored in favor of only focusing on steroidal hormones.

The results of this study indicate that while there was correlation between chemical and biological assay results in terms of trends through the plants, the bioassay results were up to an order of magnitude less than a Model of Concentration Addition would predict. This could be attributed to compounds present in the samples that have an anti-estrogenic effect, interactions between different TOrCs including competitive binding to the receptor or poor bioassay performance due to matrix interference.

In most cases, the Model of Concentration Addition showed a greater response than the YES bioassay, which is similar to findings reported by others (Rajapakse et al., 2004, Thorpe et al., 2006). Further, the measured (bioassay) vs. estimated (Model of Concentration Addition) total estrogenicity for liquid data showed better agreement than those for solids. The KBluc results were typically somewhat higher than the YES results, but still less than the individual compound predicted values.

In conclusion, this study provided a unique data set for describing instantaneous loads of hormones, AWIs, pharmaceuticals, and total estrogenicity at four WWTPs. While a major observation from the sampling program and subsequent analysis of the flows and solids loading data indicates that a comprehensive high-frequency sampling program is necessary to fully characterize mass balance and loads of estrogenicity and estrogenic compounds for any one plant, the data provided unique insights to the transfer and reduction of estrogenicity through each unit process, as well as specific instantaneous loads of non-estrogenic TOrCs of interest. This project revealed several opportunities for future research as described in the next section.

CHAPTER 5.0

RESEARCH NEEDS

This section outlines research needs associated with the estrogenicity of biosolids.

5.1 TOrC Mass Balance

A major observation from the sampling program and subsequent analysis of the flows and solids loading data indicates that a comprehensive high-frequency sampling program is necessary to fully characterize mass balance and loads of estrogenicity and TOrCs for any one plant. It is recommended that future research in this area first establishes the variability in performance on individual unit processes in which inputs, including recycle streams and sidestreams, retention times and discontinuous flows are fully characterized. Then, focused sampling at high frequency, with concommittent continuous or high frequency monitoring of plant operating parameters, such as temperature, flow, suspended solids and nutrient concentrations, can be used to evaluate whether the observed instantaneous loads, whether of pharmaceuticals, estrogenic compounds, or total estrogenicity, are indicative of plant operations during the sampling periods, or if the observed variations reflect inherent variations in samples from these complex challenging environments. Based on a more comprehensive understanding of these data, the number of sampling points needed to generate statistically relevant results can be determined from which a more reliable mass balance analysis can developed.

5.2 Chemical and Biological Assay Correlation

The results of this study indicate that while there was correlation between chemical and biological assay results in terms of trends through the plants, the bioassay results were up to an order of magnitude less than a Model of Concentration Addition would predict. This could be attributed to compounds present in the samples that have an anti-estrogenic effect, interactions between different TOrCs including competitive binding to the receptor or poor bioassay performance due to matrix interference. It is not expected that an additivity model for the summed effect of a mixture of estrogenic compounds will necessarily apply, however as bioassays are increasingly relied upon as a screening tool for estrogenicity related to wastewater treatment, it will be important to further evaluate analytical discrepancies between individual chemical and biological assay results to better quantify the actual, relevant estrogenic strength of a water or biosolids sample.

5.3 Digestion

The results of this project, as well as published studies by other researchers, suggest that the effectiveness of biosolids digestion in reducing estrogenicity and other TOrC concentrations is highly variable. However, this variation may be a function of whether the digestion is aerobic or anaerobic as well as the SRT. This is not to say that a range of other variables, such as temperature (meso- vs. thermophilic), hydraulic retention time (HRT), and C/N/P ratio, are not likely also influential, but simply that it is expected the greatest control over removal of estrogenicity will be gained by manipulation of the SRT and oxygen availability of a digester. It is with this in mind, that the team suggests possible bench-scale digester studies particularly focused on these variables and on combinations of these variables such as anaerobic followed by aerobic digestion.

5.4 Land Application

There is a lot of interest among regulators and the scientific community about the implication of TOrCs in biosolids-amended soils. Research should be conducted to develop a better understanding of the fate, transport behavior and exposure of TOrCs in biosolids-amended soil, included composted biosolids. Research should seek to address the following major knowledge gaps:

1. Mobility: including runoff and leaching evaluations.
2. Persistence: including degradation and volatilization.
3. Uptake, bioaccumulation and other factors affecting toxicity: including plants, soil biota and bioaccumulation biota predators.
4. Soil microbial impacts: including factors related to community changes and antibacterial resistance.
5. Potential to reach groundwater.
6. Short- and long-term bioavailability: including sorption and humification evaluations.
7. Validation of predictive models: including fate, transport, exposure and other factors related to effects and risk.

5.5 Lime Stabilization

Analysis of lime stabilized sludges to build on results for Plant C, which showed increases of total estrogenic activity but reductions in most hormones and significant reductions in alkylphenolic compounds. The results warrant additional work to determine whether these results are repeatable.

As stated in Section 3.3.5, lime stabilization is widely used to stabilize sludges. In this process, lime is added to untreated sludge in sufficient quantity to raise the pH to 12 or higher, which can drastically change the chemistry of many TOrCs.

The goal of research should be to first determine whether the increase in estrogenicity relative to target compound concentrations at Plant C is a real phenomenon or an analytical artifact, and then to determine if it is repeatable at other facilities utilizing similar processes. In the event that addition of lime has the effect of mobilizing hydrophobic organics that otherwise would remain sorbed to the solid phase, there would be significant implications on the mobility of these compounds to the environment after sludge disposal.

5.6 Centrate Streams

As stated in Section 2.5, centrate samples from the study plants were consistently difficult to extract and analyze for both the UA and USGS laboratories. Additionally, there were differences in ease of sample processing between the different study plants. It is possible that a colloidal phase that is not removed by centrifugation (plant or laboratory) and filtration was present in these samples. Since the plants use polymer addition to thicken and flocculate sludge it can be hypothesized that polymer may be acting as or enhancing the colloidal phase. A polymer-initiated colloidal phase that persists in the liquid phase through treatment may have significant implications for the transport of estrogens and other emerging contaminants through the treatment process, as well as the distribution of these compounds between solid and liquid phases during and after treatment.

Research should seek to use different polymers and sludges to evaluate this hypothesis. For example, a cationic polymer is added to thicken the WAS from the secondary clarifiers at Plant D; a sample of the WAS could be collected and processed with and without addition of polymer. Following centrifugation in the lab, the liquid and solid phases can be separated and analyzed for chemical constituents and estrogenicity to quantify their distribution between solids and liquids, and to assess whether polymer addition may enhance the apparent solubility of target compounds and corresponding load in the treated liquid discharge.

5.7 Secondary Treatment

Although the focus of this research was on the solids treatment process, there is a significant body of evidence that significant reductions of microconstituents take place in the biological phase of liquid wastewater treatment. Future research might seek to compare the efficiency of estrogenic activity removal during a biological nutrient removal process versus a conventional activated sludge treatment process.

REFERENCES

Aerni, H.R.; Kobler, B.; Rutishauser, B.V.; Wettstein, F.E.; Fischer, R.; Giger, W.; Hungerbüler, A.; Marazuela, M.D.; Peter, A.; Schőnenberger, R.; Vőgeli, A.C.; Suter, M.J.-F.; Effen, R.I.L., 2004. Combined biological and chemical assessment of estrogenic activities in wastewater treatment plant effluents. *Anal. Bioanal. Chem.,* 378: 688-696.

Ahel, M.; Giger, W.; Koch, M., 1994. Behaviour of Alkylphenol Polyethoxylate Surfactants in the Aquatic Environment - I. Occurrence and Transformation in Sewage Treatment, *Water Research,* 28: 1131-1142.

Burkhardt, M.R.; ReVell, R.C.; Smith, S.G.; Zaugg, S.D., 2005. Pressurized liquid extraction using water/isopropanol coupled with solid-phase extraction cleanup for industrial and anthropogenic waste-indicator compounds in sediment. *Analytica Chimica Acta,* 534: 89-100.

Burkhardt, M.R.; Zaugg, S.D.; Smith, S.G.; and ReVello, R.C., 2006. Determination of wastewater compounds in sediment and soil by pressurized solvent extraction, solid-phase extraction, and capillary-column gas chromatography/mass spectrometry: U.S. Geological Survey Techniques and Methods, book 5, chap. B2: 40.

Cahill, J.D.; Furlong, E.T.; Burkhardt, M.R., Kolpin, D.W., and Anderson, L.G., 2004. Determination of pharmaceutical compounds in surface- and ground-water samples by solid-phase extraction and high-performance liquid chromatography/electrospray ionization mass spectrometry: Journal of Chromatography A, v. 104: 171-180.

Chen, M.-Y..; Ike, M.; Fujita, M., 2002. Acute toxicity, mutagenicity, and estrogenicity of Bisphenol-A and other Bisphenols. *Environ. Toxicol.,* 17: 80-86.

De Boever, P.; Demare, W.; Vanderperren, E.; Cooreman, K.; Bossier, P.; Verstraete, W., 2001. Optimization of a yeast estrogen screen and its applicability to study the release of estrogenic isoflavones from a soygerm powder. *Environmental Health Perspectives*, 109(7): 691-697.

Dowers, T.S.; Rock, D.A.; Perkins, B.N.S.; Jones, J.P., 2004. An analysis of the regioselectivity of aromatic hydroxylation and *N*-oxygenation by cytochrome P40 enzymes. *Drug Metabolism and Disposition,* 32 (3): 328-332.

Environment Canada, Health Canada, 2001. Priority substances list assessment report: nonylphenol and its ethoxylates.

Fang, H.; Tong, W.; Perkins, R.; Soto, A.M.; Brechtl, N.V.; Sheehan, D.M., 2000. Quantitative comparisons of *in vitro* assays for estrogenic activities. *Environ. Health Perspect.,* 108: 723-729.

Folmar, L.C.; Hemmer, M.J.; Denslow, N.D.; Kroll, K.; Chen, J.; Cheek, A.; Richamn, H.; Meredith, H.; Grau, E.G., 2002. A comparison of the estrogenic potencies of estradiol, ethynylestradiol, diethystilbestrol, nonylphenol and methoxychlor in vivo and *in vitro*. *Aquatic Toxicology,* 60: 101-110.

Fraser, T.R., 1872. The antagonism between the actions of active substances. *British Medical Journal,* 2: 485-487.

Frische, T.; Faust, M.; Wiebke, M., 2009. Toxic masking and synergistic modulation of the estrogenic activity of chemical mixtures in a yeast estrogen screen (YES). *Environ. Sci. Pollut. Res.,* 16: 593-603.

Furlong, E.T.; Werner, S.L.; Anderson, B.D.; Cahill J.D., 2008. Determination of human-health pharmaceuticals in filtered water by chemically modified styrene-divinylbenzene resin-based solid-phase extraction and high-performance liquid chromatography/mass spectrometry: U.S. Geological Survey Techniques and Methods, book 5, sec. B, chap. B5: 56.

Furlong, E.T.; Gray, J.L.; Quanrud, D.M.; Teske, S.S.; Werner, S.L.; Esposito, K.; Marine, J.; Ela, W.P.; Zaugg, S.D.; Phillips, P.J.; and Stinson, B., 2010. Hormones, Pharmaceuticals, Anthropogenic Waste Indicators, and Total Estrogenicity in Liquid and Solid Samples Collected to Estimate the Fate of Estrogenic Compounds During Municipal Sludge Stabilization and Dewatering, U.S. Geological Survey Data Report, in review.

Glassmeyer, S.; Kolpin, D.W.; Furlong, E.T.; Focazio, M.T., 2007. Environmental presence and persistence of pharmaceuticals: An overview: in Fate of Pharmaceuticals in the Environment and in Water Treatment Systems, Diana S. Aga (ed), CRC Press, Taylor and Francis Books. Boca Raton, FL, p408.

Gray, J.L.; Foreman, W.T.; Lindley, C.E.; Revello R.C.; Barber, L.B., 2010. Determination of Steroid Hormones in Filtered and Unfiltered Water by Solid-Phase Extraction, Derivatization and Gas Chromatography with Tandem Mass Spectrometry, U.S. Geological Survey Techniques and Methods, in preparation.

GWRC, 2008. Tools to detect estrogenicity in environmental waters – Final report. Global Water Research Coalition (GWRC) / Water Environment Research Foundation (WERF).

Johnson, A.C.; Sumpter, J.P., 2001. Removal of endocrine-disrupting chemicals in activated sludge treatment works. *Environmental Science and Technology,* 35 (24): 4697-4703.

Joss et al., 2004. Removal of Estrogens in Municipal Wastewater Treatment under Aerobic and Anaerobic Conditions: Consequences for Plant Optimization. *Environmental Science and Technology,* 38: 3047-3055.

Kawamura, Y.; Ogawa, Y.; Nishimura, T.; Kikuchi, Y.; Nishikawa, J.; Nishihara, T.; Tanamoto, K., 2003. Estrogenic activities of UV stabilizers used in food contact plastics and benzophenone derivatives tested by the yeast two-hybrid assay. *Journal of Health Science,* 49 (3): 205-212.

Kinney, C.A.; Furlong, E.T.; Werner, S.L.; Cahill, J.D., 2006a. Presence and Distribution of Wastewater-Derived Pharmaceuticals in Soil Irrigated With Reclaimed Water. *Environmental Toxicology and Chemistry*, 25(2), 317-326.

Kinney, C.A.; Furlong, E.T.; Zaugg, S.D.; Burkhardt, M.R.; Werner, S.L.; Cahill, J.D.; Jorgensen, G.R., 2006b. Survey of Organic Wastewater Contaminants in Biosolids Destined for Land Application. *Environmental Science and Technology,* 40(23): 7207-7215.

Kitamura, S.; Sugihara, K.; Sanoh, S.; Fujimoto, N.; Ohta, S., 2008. Metabolic activation of proestrogens in the environment by cytorchrome P450 system. *Journal of Health Science,* 54 (4): 343-355.

Kolpin, D.W.; Furlong, E.T.; Meyer, M.T.; Thurman, E.M.; Zaugg, S.D.; Barber, L.B.; Buxton, H.T., 2002. Pharmaceuticals, hormones, and other organic wastewater contaminants in US streams, 1999-2000: A national reconnaissance: *Environmental Science and Technology,* 36(6): 1202-1211.

Kunz, P.Y.; Fent, K., 2006. Estrogenic activity of UV filter mixtures. *Toxicology and Applied Pharmacology,* 217: 86-99.

Liu, Z.-h.; Kanjo, Y.; Mizutani, S., 2009. Removal mechanisms for endocrine disrupting compounds (EDCs) in wastewater treatment - physical means, biodegradation, and chemical advanced oxidation: A review. *Science of the Total Environment*, 407 (2): 731-748.

Loewe, S.; Muischnek, H.,1926. Über kombinationswirkungen. 1. Mitteliung: Hilfsmittel der Fragestellung. Naunyn-Schmiedebergs *Arch. Exp. Pathol. Pharmokol.*, 114: 313-326.

Matsumoto, T.; Kobayashi, M.; Moriwaki, T.; Kawai, S.; Watabe, S., 2004. Survey of estrogenic activitiy in fish feed by yeast estrogen-screen assay. *Comparative Biochemistry and Physiology,* Part C 139: 147-152.

Metcalf and Eddy, 2003. *Wastewater Engineering: Treatment and Reuse*. McGraw-Hill. New York, NY.

Miller, D.; Wheals, B.B.; Beresford, N.; Sumpter, J.P., 2001. Estrogenic Activity of Phenolic Additives Determined by an *In vitro* Yeast Bioassay. *Environmental Health Perspectives,* 109 (2): 133-138.

Okubo, T.; Suzuki, T.; Yokoyama, Y.; Kano, K.;, Kano, I., 2003. Estimation of estrogenic and anti-estrogenic activities of some phthalate diesters and monoesters by MDF-7 cell proliferation assay *in vitro*. *Biol. Pharm. Bull.,* 26 (8): 1219-1224.

Petrovic, M.; Eljarrat, E.; Lopez de Alda, M.J.; Barcelo, D., 2004. Endocrine disrupting compounds and other emerging contaminants in the environment: A survey on new monitoring strategies and occurrence data. *Anal. Bioanal. Chem.,* 378: 549-562.

Rajapakse, N.; Ong, D.; Kortenkamp, A., 2001. Defining the impact of weakly estrogenic chemicals on the action of steroidal estrogens. *Toxicological Sciences,* 60: 296-304.

Rajapakse, N.; Silva, E.; Scholze, M.; Kortenkamp, A., 2004. Deviation from additivity with estrogenic mixtures containing 4-nonylphenol and 4-*tert*-octylphenol detected in the E-Screen Assay. *Environmental Science and Technology,* 38: 6343-6352.

Routledge, E.J.; Sumpter, J.P., 1996. Estrogenic activity of surfactants and some of their degradation products assessed using a recombinant yeast screen. *Environmental Toxicology and Chemistry,* 15(3): 241-248.

Routledge, E.J.; Sheahan, D.; Desbrow, C.; Brighty, G.C.; Waldock, M.; Sumpter, J.P., 1998. Identification of estrogenic chemicals in STW effluent. 2. *In vivo* responses in trout and roach. *Environ. Sci. Tech.,* 32(11): 1559-1565.

Sanseverino, J.; Gupta, R.K.; Layton, A.C.; Patterson, S.S.; Ripp, S.A.; Saidak, L.; Simpson, M.L.; Schultz, T.W.; Sayler, G.S., 2005. Use of *Saccharomyces cerevisiae* BLYES expressing bacterial bioluminescence for rapid, sensitive detection of estrogenic compounds. *Applied and Environmental Microbiology,* 71 (8): 4455-4460.

Sheahan, D.A.; Brighty, G.C.; Daniel, M.; Jobling, S.; Harries, J.E., et al., 2002. Reduction in the Estrogenic Activity of a Treated Sewage Effluent Discharge to an English River as a Result of a Decrease in the Concentration of Industrially Derived Surfactants. Environmental Toxicology and Chemistry, 21(3): 515-519.

Silva, E.; Rajapakse, N.; Kortenkamp, A., (2002). Something from "nothing"– eight weak estrogenic chemicals combined at concentrations below NOECs produce significant mixture effects. *Environmental Science and Technology,* 36: 1751-1756.

Snyder, S.A.; Westerhoff, P.; Yoon, Y.; Sedlak, D.L., 2003, Pharmaceuticals, personal care products, and endocrine disruptors in water: Implications for the water industry. *Environmental Engineering Science*, 20 (5): 449-469.

Soto, A.M.; Sonnenschein, C.; Chung, K.L.; Fernandez M.F.; Olea, N.; Serrano, F.O., 1995. The E-screen assay as a tool to identify estrogens: an update on estrogenic environmental pollutants. *Environ. Health Perspect.,* 103: 113-122.

Sumpter, J.P., 2005. Endocrine disrupters in the aquatic environment: An overview, *Acta Hydrochimica et Hydrobiologica*, 33 (1): 9-16.

Ternes et al., 1999. Behavior and occurrence of estrogens in municipal sewage treatment plants I. – Investigations in Germany, Canada and Brazil. *The Science of the Total Environment,* 225: 81-90.

Ternes et al., 1999. Behaviour and occurrence of estrogens in municipal sewage treatment plants – II. Aerobic batch experiments with activated sludge. *The Science of the Total Environment,* 225: 91-99.

Thorpe, K., et al., 2006. An assessment of the model of concentration addition for predicting the estrogenic activity of chemical mixtures in wastewater treatment works effluents. *Environmental Health Perspectives*, 114: 90-97.

U.S. EPA, 1981. Land Application of Municipal Sewage Sludge for the Production of Fruits and VegeTables: A Statement of Federal Policy and Guidance. SW 905 U.S. Environmental Protection Agency, U.S. Food and Drug Administration, and U.S. Department of Agriculture, Washington, D.C.

U.S. EPA, 1989. Preparing perfect project plans – pocket guide for preparation of quality assurance project plans: U.S. Environmental Protection Agency Report EPA/600/0-89/087: 62.

U.S. EPA, 1991. Interagency Policy on Beneficial Use of Municipal Sewage Sludge on Federal Land; Notice. *Federal Register,* 56(138): 33186-33188.

U.S. EPA, 1999. Biosolids Generation, Use,and Disposal in The United States, EPA530-R-99-009, Municipal and Industrial Solid Waste Division, Office of Solid Waste, September 1999.

U.S. EPA, 2005. Guidelines establishing test procedures for the analysis of pollutants (App. B, Part 136, Definition and procedures for the determination of the method detection limit): U.S. Code of Federal Regulations, Title 40, revised as of July 1, 2005: 319–322

USGS, 1997-1999. National field manual for the collection of water-quality data: U.S. Geological Survey Techniques of Water-Resources Investigations, book 9, chaps. A1-A9, 2 v., variously paged. Available online: *http://pubs.water.usgs.gov/twri9.* Chapters were published from 1997-1999; updates and revisions are ongoing and can be viewed at: *http://water.usgs.gov/owq/FieldManual/mastererrata.html.*

USGS, 2009. National field manual for the collection of water-quality data: U.S. Geological Survey Techniques of Water-Resources Investigations, book 9, chaps. A1-A9, available online at http://pubs.water.usgs.gov/twri9A, last accessed November, 23, 2009.

Vajda, A.M.; Barber, L.B.; Gray, J.L.; Lopez, E.M.; Woodling, J.O.; Norris, D.O., 2008. Reproductive Disruption in Fish Downstream from an Estrogenic Wastewater Effluent. *Environ. Sci. Technol.*, 42(9): 3407-3414.

Van Den Heuvel, M.R.; Leusch, F.D.L.; Taylor, S.; Shannon, N.; McKague, A.B., 2006. Assessment of the reproductive-endocrine disrupting potential of chlorine dioxide oxidation products of plant sterols. *Environ. Sci. Tech.,* 40(8): 2594-2600.

Wilson, V.S.; Bobseine, K.; Gray Jr., L.E., 2004. Development and characterization of a cell line that stably expresses an estrogen-responsive luciferase reporter for the detection of estrogen receptor agonist and antagonists. *Toxicol. Sciences*, 81: 69-77.

Zaugg, S.D.; Smith, S.G.; Schroeder, M.P.; Barber, L.B.; and Burkhardt, M.R., 2002. Methods of analysis by the U.S. Geological Survey National Water Quality Laboratory – Determination of wastewater compounds by polystyrene-divinylbenzene solid-phase extraction and capillary-column gas chromatography/mass spectrometry: U.S. Geological Survey Water-Resources Investigations Report 01-4186: 37.

Zoller, U., 2009. Chapter 3.6 Alkylphenol Ethoxylates and their Raw Materials, Handbook of Detergents Part F: Production, surfactant science series volume 142: 61-62.

WASTEWATER UTILITY

Alabama
Montgomery Water Works & Sanitary Sewer Board

Alaska
Anchorage Water & Wastewater Utility

Arizona
Avondale, City of
Glendale, City of, Utilities Department
Mesa, City of
Peoria, City of
Phoenix Water Services Dept.
Pima County Wastewater Management
Safford, City of
Tempe, City of

Arkansas
Little Rock Wastewater Utility

Califomia
Central Contra Costa Sanitary District
Corona, City of
Crestline Sanitation District
Delta Diablo Sanitation District
Dublin San Ramon Services District
East Bay Dischargers Authority
East Bay Municipal Utility District
El Dorado Irrigation District
Fairfield-Suisun Sewer District
Fresno Department of Public Utilities
Inland Empire Utilities Agency
Irvine Ranch Water District
Las Gallinas Valley Sanitary District
Las Virgenes Municipal Water District
Livermore, City of
Los Angeles, City of
Los Angeles County, Sanitation Districts of
Napa Sanitation District
Novato Sanitary District
Orange County Sanitation District
Palo Alto, City of
Riverside, City of
Sacramento Regional County Sanitation District
San Diego Metropolitan Wastewater Department, City of
San Francisco, City & County of
San Jose, City of
Santa Barbara, City of
Santa Cruz, City of
Santa Rosa, City of
South Bayside System Authority
South Coast Water District
South Orange County Wastewater Authority
South Tahoe Public Utility District
Stege Sanitary District
Sunnyvale, City of
Union Sanitary District
West Valley Sanitation District

Colorado
Aurora, City of
Boulder, City of
Greeley, City of
Littleton/Englewood Water Pollution Control Plant
Metro Wastewater Reclamation District, Denver

Connecticut
Greater New Haven WPCA
Stamford, City of

District of Columbia
District of Columbia Water & Sewer Authority

Florida
Broward, County of
Fort Lauderdale, City of
Jacksonville Electric Authority (JEA)
Miami-Dade Water & Sewer Authority
Orange County Utilities Department
Pinellas, County of
Reedy Creek Improvement District
Seminole County Environmental Services
St. Petersburg, City of
Tallahassee, City of
Toho Water Authority
West Palm Beach, City of

Georgia
Atlanta Department of Watershed Management
Augusta, City of
Clayton County Water Authority
Cobb County Water System
Columbus Water Works
Fulton County
Gwinnett County Department of Public Utilities
Savannah, City of

Hawaii
Honolulu, City & County of

Idaho
Boise, City of

Illinois
Decatur, Sanitary District of
Greater Peoria Sanitary District
Kankakee River Metropolitan Agency
Metropolitan Water Reclamation District of Greater Chicago
Wheaton Sanitary District

Indiana
Jeffersonville, City of

Iowa
Ames, City of
Cedar Rapids Wastewater Facility
Des Moines, City of
Iowa City

Kansas
Johnson County Wastewater
Unified Government of Wyandotte County/ Kansas City, City of

Kentucky
Louisville & Jefferson County Metropolitan Sewer District
Sanitation District No. 1

Louisiana
Sewerage & Water Board of New Orleans

Maine
Bangor, City of
Portland Water District

Maryland
Anne Arundel County Bureau of Utility Operations
Howard County Bureau of Utilities
Washington Suburban Sanitary Commission

Massachusetts
Boston Water & Sewer Commission
Massachusetts Water Resources Authority (MWRA)
Upper Blackstone Water Pollution Abatement District

Michigan
Ann Arbor, City of
Detroit, City of
Holland Board of Public Works
Saginaw, City of
Wayne County Department of Environment
Wyoming, City of

Minnesota
Rochester, City of
Western Lake Superior Sanitary District

Missouri
Independence, City of
Kansas City Missouri Water Services Department
Little Blue Valley Sewer District
Metropolitan St. Louis Sewer District

Nebraska
Lincoln Wastewater & Solid Waste System

Nevada
Henderson, City of
Las Vegas, City of
Reno, City of

New Jersey
Bergen County Utilities Authority
Ocean County Utilities Authority

New York
New York City Department of Environmental Protection

North Carolina
Charlotte/Mecklenburg Utilities
Durham, City of
Metropolitan Sewerage District of Buncombe County
Orange Water & Sewer Authority
University of North Carolina, Chapel Hill

Ohio
Akron, City of
Butler County Department of Environmental Services
Columbus, City of
Metropolitan Sewer District of Greater Cincinnati
Montgomery, County of
Northeast Ohio Regional Sewer District
Summit, County of

Oklahoma
Oklahoma City Water & Wastewater Utility Department
Tulsa, City of

Oregon
Albany, City of
Clean Water Services
Eugene, City of
Gresham, City of
Portland, City of Bureau of Environmental Services
Lake Oswego, City of
Oak Lodge Sanitary District
Water Environment Services

Pennsylvania
Hemlock Municipal Sewer Cooperative (HMSC)
Philadelphia, City of
University Area Joint Authority

South Carolina
Charleston Water System
Mount Pleasant Waterworks & Sewer Commission
Spartanburg Water

Tennessee
Cleveland Utilities
Murfreesboro Water & Sewer Department
Nashville Metro Water Services

Texas
Austin, City of
Dallas Water Utilities
Denton, City of
El Paso Water Utilities

Fort Worth, City of
Houston, City of
San Antonio Water System
Trinity River Authority

Utah
Salt Lake City Corporation

Virginia
Alexandria Sanitation Authority
Arlington, County of
Fairfax, County of
Hampton Roads Sanitation District
Hanover, County of
Henrico, County of
Hopewell Regional Wastewater Treatment Facility
Loudoun Water
Lynchburg Regional Wastewater Treatment Plant
Prince William County Service Authority
Richmond, City of
Rivanna Water & Sewer Authority

Washington
Everett, City of
King County Department of Natural Resources
Seattle Public Utilities
Sunnyside, Port of
Yakima, City of

Wisconsin
Green Bay Metro Sewerage District
Kenosha Water Utility
Madison Metropolitan Sewerage District
Milwaukee Metropolitan Sewerage District
Racine, City of
Sheboygan Regional Wastewater Treatment
Wausau Water Works

Water Services Association of Australia
ACTEW Corporation
Barwon Water
Central Highlands Water
City West Water
Coliban Water Corporation
Cradle Mountain Water
Gippsland Water
Gladstone Area Water Board
Gold Coast Water
Gosford City Council
Hunter Water Corporation
Logan Water
Melbourne Water
Moreton Bay Water
Onstream
Power & Water Corporation
Queensland Urban Utilities
SEQ Water
South Australia Water Corporation
Sunshine Coast Water
Sydney Catchment Authority
Sydney Water
Unity Water
Wannon Regional Water Corporation
Watercare Services Limited (NZ)
Water Corporation
Western Water
Yarra Valley Water

Canada
Edmonton, City of/Edmonton Waste Management Centre of Excellence
Lethbridge, City of
Regina, City of, Saskatchewan
Toronto, City of, Ontario
Winnipeg, City of, Manitoba

STORMWATER UTILITY

California
Fresno Metropolitan Flood Control District
Los Angeles, City of, Department of Public Works
Monterey, City of
San Francisco, City & County of
Santa Rosa, City of
Sunnyvale, City of

Colorado
Aurora, City of
Boulder, City of

Florida
Orlando, City of

Iowa
Cedar Rapids Wastewater Facility
Des Moines, City of

Kansas
Lenexa, City of
Overland Park, City of

Kentucky
Louisville & Jefferson County Metropolitan Sewer District

Maine
Portland Water District

North Carolina
Charlotte, City of, Stormwater Services

Pennsylvania
Philadelphia, City of

Tennessee
Chattanooga Stormwater Management

Texas
Harris County Flood Control District, Texas

Washington
Bellevue Utilities Department
Seattle Public Utilities

STATE

Connecticut Department of Environmental Protection
Kansas Department of Health & Environment
New England Interstate Water Pollution Control Commission (NEIWPCC)
Ohio Environmental Protection Agency
Ohio River Valley Sanitation Commission
Urban Drainage & Flood Control District, CO

CORPORATE

ADS LLC
Advanced Data Mining International
AECOM
Alan Plummer & Associates
Alpine Technology Inc.
Aqua-Aerobic Systems Inc.
Aquateam–Norwegian Water Technology Centre A/S
ARCADIS
Associated Engineering
Bernardin Lochmueller & Associates
Black & Veatch
Blue Water Technologies, Inc.
Brown & Caldwell
Burgess & Niple, Ltd.
Burns & McDonnell
CABE Associates Inc.
The Cadmus Group
Camp Dresser & McKee Inc.
Carollo Engineers Inc.
Carpenter Environmental Associates Inc.
CET Engineering Services
CH2M HILL
CRA Infrastructure & Engineering
CONTECH Stormwater Solutions
D&B/Guarino Engineers, LLC
Damon S. Williams Associates, LLC
Ecovation
EMA Inc.
Environmental Operating Solutions, Inc.
Environ International Corporation
Fay, Spofford, & Thorndike Inc.
Freese & Nichols, Inc.
ftn Associates Inc.
Gannett Fleming Inc.
Garden & Associates, Ltd.
Geosyntec Consultants
GHD Inc.
Global Water Associates
Greeley and Hansen LLC
Hazen & Sawyer, P.C.
HDR Engineering Inc.
HNTB Corporation
Hydromantis Inc.
HydroQual Inc.
Infilco Degremont Inc.
Jason Consultants LLC Inc.
Jordan, Jones, & Goulding Inc.
KCI Technologies Inc.
Kelly & Weaver, P.C.
Kennedy/Jenks Consultants
Larry Walker Associates
LimnoTech Inc.
Lombardo Associates, Inc.
The Low Impact Development Center Inc.
Malcolm Pirnie Inc.
Material Matters, Inc.
McKim & Creed
MWH
NTL Alaska, Inc.
O'Brien & Gere Engineers Inc.
Odor & Corrosion Technology Consultants Inc.
Parametrix Inc.
Parsons
Post, Buckley, Schuh & Jernigan
Praxair, Inc.
RMC Water & Environment
Ross & Associates Ltd.
SAIC
Siemens Water Technologies
The Soap & Detergent Association
Smith & Loveless, Inc.
Southeast Environmental Engineering, LLC
Stone Environmental Inc.
Stratus Consulting Inc.
Synagro Technologies Inc.
Tetra Tech Inc.
Trojan Technologies Inc.
Trussell Technologies, Inc.
URS Corporation
Wallingford Software
Westin Engineering Inc.
Wright Water Engineers
Zoeller Pump Company

INDUSTRY

American Electric Power
American Water
Anglian Water Services, Ltd.
Chevron Energy Technology
The Coca-Cola Company
Dow Chemical Company
DuPont Company
Eastman Chemical Company
Eli Lilly & Company
InsinkErator
Johnson & Johnson
Merck & Company Inc.
Procter & Gamble Company
Suez Environnment
United Utilities North West (UUNW)
United Water Services LLC
Veolia Water North America

Note: List as of 4/20/10

IWA Publishing's authorised EU representative for General Product Safety Regulations is Diane D'Arras, 15 rue Duret, 75116 Paris, France, e-mail: safety@iwap.co.uk.

Printed and bound by CPI Group (UK) Ltd, Croydon, CR0 4YY
12/05/2026
02108868-0016